Random Telegraph Signals in Semiconductor Devices

Random Telegraph Signals in Semiconductor Devices

Eddy Simoen
Imec, Leuven, Belgium

Cor Claeys
Imec, Leuven, Belgium

IOP Publishing, Bristol, UK

ISBN 978-0-7503-1272-1 (ebook)
ISBN 978-0-7503-1273-8 (print)
ISBN 978-0-7503-1274-5 (mobi)

DOI 10.1088/978-0-7503-1272-1

Version: 20161001

IOP Expanding Physics
ISSN 2053-2563 (online)
ISSN 2054-7315 (print)

British Library Cataloguing-in-Publication Data: A catalogue record for this book is available from the British Library.

Published by IOP Publishing, wholly owned by The Institute of Physics, London

IOP Publishing, Temple Circus, Temple Way, Bristol, BS1 6HG, UK

US Office: IOP Publishing, Inc., 190 North Independence Mall West, Suite 601, Philadelphia, PA 19106, USA

Contents

Preface

The fundamental importance of random telegraph signal (RTS) in submicrometer-area, scaled metal–oxide–semiconductor (MOS) transistors was realized soon after their discovery and led to numerous publications in the field. This is related to the fact that RTS offers a unique opportunity to study single electrically active defects in a semiconductor device structure, enabling not only the unraveling of its properties (energy level and capture cross section), but also its physical position, with high spatial resolution. The recent efforts to explain the electrical oxide-trap properties from *ab initio* density functional theory calculations raise the hope that even a chemical and structural identification of the responsible traps is within reach. In addition, as RTS is believed to be the basic component of $1/f$ noise, its discovery and study led in the early 1990s to a more refined correlated mobility number fluctuation theory for $1/f$ noise in large-area MOS transistors, which is still the state-of-the-art model.

At the same time, the fact that the RTS phenomenon becomes more important for scaled devices and technologies raised practical concerns with respect to memory operation and reliability about a decade ago. This triggered the interest of the microelectronics industry and led to a boost in publications on RTS beyond the purely device physics community. RTS is still considered an important source of dynamic variability in today's complementary MOS (CMOS) technologies, and should be accounted for in both device and circuit operation. This has led to strategies to mitigate or suppress the phenomenon, either based on process/design optimization or by developing optimal circuit operation schemes. In addition to being a root cause of $1/f$ noise, it has been shown in the past decade that RTS is intimately connected to the reliability of scaled CMOS transistors, namely, to the so-called bias-temperature instability. The study of RTS has also caused a paradigm shift in the concept of reliability from a purely deterministic description to a stochastic one.

In more recent years, RTS has also been observed in devices beyond CMOS, such as carbon nanotube transistors, tunnel field-effect transistors, novel metal–oxide based resistive random access memories, etc, and it is also expected to be prominent in other future nanometric devices because of the nature of the phenomenon: scaled transistors will become more sensitive to the presence of single defects or impurities, which are beyond today's technological control. In fact, one may even hope to benefit from RTS, not only for fundamental defect studies, but perhaps to develop single-atom switches or memories.

Surprisingly, so far no textbook has been prepared on the topic of RTS in semiconductor devices—only some excellent review papers in journals. Therefore, we feel that there is a need for a more extensive and comprehensive work on this topic. It is the intention of the book to provide a detailed description of the physics and modeling of RTS in semiconductor devices and circuits. At the same time, it is assumed that the reader has a solid background in semiconductor device physics (MOS transistors, p–n junctions). A basic knowledge of semiconductor processing is also useful. Whenever necessary, references to relevant textbooks or review papers will be provided. For more novel types of devices, a brief description of the underlying physics and operation principles will be given, without becoming exhaustive, since the focus of this book is on the RTS phenomenon.

We hope that this book will contribute to a better understanding of RTS in semiconductor devices and may trigger new ideas and research in the field.

Acknowledgments

Many colleagues contributed over the years to our work on low-frequency noise and RTS studies. In particular, we would like to thank the contributions of H Achour, M Aoulaiche, B Dierickx, Carin, B Cretu, M G C de Andrade, S D dos Santos, W Fang, P Fazan, N P Garbar, W Guo, N Horiguchi, M Jurczak , B Kaczer, J W Lee, J Luo, U Magnussen, J A Martino, A Mercha, M V Petrichuk, J Sikula, M Toledano Luque, G Van den Bosch, P Vasina, C Zhao.

Author biographies

Eddy Simoen

Eddy Simoen obtained his bachelor's (1976–78) and master's degrees (1978–80) in physics engineering, and his PhD in engineering (1985) from Ghent University (Belgium). He is currently a specialist at IMEC (Leuven, Belgium), involved in the study of defect and strain engineering in high mobility and epitaxial substrates, and defect studies in germanium and III–V compounds (AlN, GaN, InP, etc). Another current focus is the study of one-transistor memories based on bulk FinFET and ultra-thin buried oxide silicon-on-insulators, using low-frequency noise. In 2013 Simoen was nominated as a part-time professor at Ghent University in the field of study of the impact of defects on semiconductor devices. He is a member of IEEE and ECS, and became an ECS Fellow in 2016. In these fields he has (co-)authored over 1500 journal and conference papers and, in addition, 12 book chapters and a monograph on *Radiation Effects in Advanced Semiconductor Devices and Materials* (2002, Springer), the Chinese translation of which was published in 2008. He was also a co-editor of the book *Germanium-based Technologies—From Materials to Devices* (2007, Elsevier; Chinese translation 2010). Another book, the *Fundamental and Technological Aspects of Extended Defects in Germanium*, was published by Springer 2009. Simoen is also the co-inventor of two patents.

Cor Claeys

Cor Claeys received his PhD from KU Leuven in Belgium, where he has been a professor since 1990. At IMEC he is currently Director of Advanced Semiconductor Technologies responsible for strategic relations. His main interests are silicon technology, device physics, low-frequency noise phenomena, radiation effects, defect engineering and material characterization. He has co-edited the books *Low Temperature Electronics* (2001, Elsevier) and *Germanium-Based Technologies—From Materials to Devices* (2007, Elsevier) and has written monographs on *Radiation Effects in Advanced Semiconductor Materials and Devices* (2002, Springer) and *Fundamental and Technological Aspects of Extended Defects in Germanium* (2009, Springer). He has authored and co-authored 14 book chapters, over 1000 conference presentations and more than 1200 technical papers. He has also been the editor or co-editor of 60 conference proceedings. Claeys is a fellow of the Electrochemical Society and of IEEE. He was the founder of the IEEE Electron Devices Benelux Chapter, Chair of the IEEE Benelux Section, an elected AdCom Member of the Electron Devices

Society and the EDS Vice President for Chapters and Regions. He was EDS President in 2008–9 and the Division Director on the IEEE Board of Directors in 2012–13. Claeys is a recipient of the IEEE Third Millennium Medal and in 2013 he received the IEEE EDS Distinguished Service Award. Within the Electrochemical Society he was Chair of the Electronics & Photonics Division from 2001 to 2003, and in 2004 received the Electronics & Photonics Division Award.

List of symbols

A	m^2	Area
A_m	Vm	Dynamic matching parameter
A_{RTS}		Normalized RTS amplitude
b	m^{-1}	Variation parameter in the electron wave-function
C_D	$F\ m^{-2}$	Depletion layer capacitance per unit area
C_G	$F\ m^{-2}$	Gate capacitance per unit area
C_{inv}	$F\ m^{-2}$	Inversion layer capacitance per unit area
C_{ox}	$F\ m^{-2}$	Oxide capacitance per unit area
d	m	Tunnel depth
d_{CNT}	m	Diameter of the carbon nanotube device
D_{it}	$m^{-2}\ eV^{-1}$	Density of interface traps
d_{NW}	m	Diameter of the nanowire
D_T	m^{-2}	Trap density in de the depletion layer
E_a	eV	Activation energy
E_b	eV	Binding energy of a carrier in a trap
E_B	eV	Energy barrier for carrier capture
E_C	eV	Conduction band minimum
E_{cap}	eV	Activation energy for capturing
E_{Coul}	eV	Coulomb energy
E_{Cox}	eV	Conduction band minimum in oxide
E_{CT}	eV	Gibbs free energy
E_{em}	eV	Activation energy for emission
E_{Fn}	eV	Quasi Fermi level for electrons
E_{Fp}	eV	Quasi Fermi level for holes
E_F	eV	Surface Fermi level
E_G	eV	Band gap energy
E_{GR}	eV	Activation energy to create a hole at the interface in silicon
E_{hop}	eV	Energy barrier for electron hopping
E_{relax}	eV	Energy related to the displacement of lattice atoms
E_T	eV	Energy level in the band gap
E_V	eV	Valence band maximum
E_Z	eV	Zeeman energy
E_0	eV	Electronic part of the energy due to a trapping event
f	Hz	Frequency
f_{max}	Hz	Maximum integration frequency
f_{min}	Hz	Minimum integration frequency
f_T		Trap occupancy by an electron
F	$V\ m^{-1}$	Electric field
F_{ox}	$V\ M^{-1}$	Oxide electric field
F_s	$V\ m^{-1}$	Surface electric field
f_T		Trap occupancy by an electron
g		Trap degeneracy factor
g_m	Ω^{-1}	Transconductance

G	Ω^{-1}	Conductance
G_D	Ω^{-1}	Drain conductance
h	Js	Planck's constant
H_{CT}	eV	Enthalpy of a trap
H_{Fin}	m	Fin height
I_B	A	Substrate current
$I_{dir\text{-}tunn}$	A	Direct-tunneling current
I_{EDT}	A	Gate edge tunnel current
I_d	A	Displacement current
I_D	A	Drain current
I_{DT}	A	Direct tunnel current
I_G	A	Gate current
I_{GIDL}	A	Gate-induced drain leakage current
I_{loc}	A	Local current around a trap
I_{HRS}	A	Current in the high resistance state
I_{READ}	A	Read current in memory cell
I_{sat}	A	Saturation current
I_{SILC}	A	Stress-induced leakage current
I_{TE}	A	Current top electrode
k	$J\ K^{-1}$	Boltzmann constant
L	m	Gate length
L_C	m	Screening length by the 2DEG
L_{eff}	m	Effective gate length
L_t	m	Trap length of a square-shaped trap area
L_{to}	m	Trap length of a charge square-shaped trap area
m^*_e	kg	Electron effective mass
m^*_h	kg	Hole effective mass
m_l	kg	Longitudinal effective mass
m_{ox}	kg	Electron effective mass in SiO_2
m_t	kg	Transversal effective mass
m_{var}	A	Average value of a variable
n	m^{-3}	Carrier concentration
N	m^{-3}	Number of carriers in the channel
N_a	m^{-3}	Substrate doping concentration (nMOS)
N_C	m^{-3}	Density of states in the conduction band
N_D	m^{-3}	Doping density
N_{ot}	m^{-2}	Density of oxide traps
N_{RTS}		Number of RTSs
n_s	m^{-2}	Surface carrier concentration
n_t		Number of traps
N_t	m^{-3}	Density of traps
N_V	m^{-3}	Density of states in the valence band
P_C		Capture probability
P_{etun}		Capture probability by tunneling
P_{htun}		Hole tunneling probability
q	C	Electron charge

q^*	C	Effective trap charge including Coulomb repulsion
Q_B	C cm^{-2}	Bulk charge density
Q_G	C	Image charge
Q_{inv}	C m^{-2}	Inversion charge density
Q_n	C m^{-2}	Channel charge density
r	m	Distance from a trap
r_0	m	Distance between two traps
R	m	Radius of a circular device
R_c		Coupling coefficient
R_D	Ω	Channel resistance
R_e	s^{-1}	Capture rate of an electron
R_{gen}	s^{-1}	Carrier generation rate
R_h	s^{-1}	Capture rate of a hole
R_Q	m	Radial charge position
R_S	Ω	Channel resistance
R_{thr}	s^{-1}	Rate of thermal excitation
s		Trap polarity
S	J K^{-1}	Entropy
S_0		Huang Rhys factor related to the lattice coupling
S_{IG}	A^2 Hz^{-1}	Gate-current noise spectral density
SS	V/dec	Subthreshold swing
S_{VG}	V^2 Hz^{-1}	Input-referred voltage noise spectral density
t	s	Time
T	K	Temperature
t_b	m	Barrier thickness
t_{dp}	m	Thickness of polysilicon depletion layer
t_{dp}	m	Polysilicon depletion
T_e	K	Electron temperature
t_h	s	Time in the RTS high state
t_{HK}	m	High-κ dielectric layer thickness
t_{IL}	m	Interfacial layer thickness
t_{inv}	m	Inversion layer thickness
t_l	s	Time in the RTS low state
t_{ox}	m	Oxide thickness
t_{qm}	m	Carriers centroid offset due to inversion layer quantum effects
t_{relax}	s	Relaxation time
t_{rise}	s	Rise time
t_{RW}	s	Period of random walk RTS
t_s	s	Sampling time
t_{stress}	s	Stress time
V_{BS}	V	Substrate voltage
V_c	V	Voltage difference between the quasi Fermi levels
V_{DS}	V	Drain voltage
V_{DSFmax}	V	Drain voltage for maximum τ_c/τ_e ratio in forward operation
V_{DSRmax}	V	Drain voltage for maximum τ_c/τ_e ratio in reverse operation
V_{FB}	V	Flat-band voltage

V_{GS}	V	Gate voltage
$V_{GS\text{-back}}$	V	Gate voltage of the back transistor
$V_{GS\text{-front}}$	V	Gate voltage of the front transistor
V_{GS0}	V	Gate voltage where $\tau_c = \tau_e$
v_n	m s^{-1}	Carrier velocity
V_{oxe}	V	Potential barrier for an electron from Si to SiO_2
V_{oxh}	V	Potential barrier for a hole from Si to SiO_2
V_{read}	V	Read voltage
V_{relax}	V	Relaxation gate voltage
V_{reset}	V	Reset voltage
V_{stress}	V	Stress gate voltage
V_T	V	Threshold voltage
V_{TE}	V	Voltage top electrode
v_{thn}	m s^{-1}	Thermal velocity for electrons
v_{thp}	m s^{-1}	Thermal velocity for holes
w	m	Width depletion layer
W	m	Gate width
W_{eff}	m	Effective gate width
W_{Fin}	m	Fin width
W_t	m	Width of a trap
x_T	m	Trap depth from the interface
x_T^{CL}	m	Trap depth from the interface (classical calculation)
x_T^{QM}	m	Trap depth from the interface (quantum mechanical calculation)
y_T	m	Lateral trap position in the channel
Z_{eff}	m	Effective trap position in the dielectric in the HRS

List of Greek symbols

α_c	V^{-1}	Coupling factor between capture time constant and gate voltage
α_e	V^{-1}	Coupling factor between emission time constant and gate voltage
α_{sc}	Vs	Scattering coefficient
α_t	m	Tunneling parameter
γ_c	s^{-1}	Tunnel rate for capture
γ_e	s^{-1}	Tunnel rate for emission
γ_{1s}		Percolation area coefficient in subthreshold regime
γ_1		Percolation area coefficient in inversion regime
Δ		Tunneling matrix element
ΔI_D	A	RTS amplitude
ε_0	F cm^{-1}	Permittivity of the free space
ε_{eff}	F cm^{-1}	Effective permittivity
ε_{HK}		Dielectric constant of high-κ dielectric
ε_{IL}		Dielectric constant of interfacial layer
ε_{ox}		Dielectric constant of SiO_2
ε_{sem}		Dielectric constant of the semiconductor
ε_{Si}		Dielectric constant of silicon
η_{BOX}	V	RTS voltage amplitude in buried oxide
η_{GOX}	V	RTS voltage amplitude in (front) gate oxide
η_o	V	RTS voltage amplitude (before electrical stress)
η_{stress}	V	RTS voltage amplitude (after electrical stress)
κ		Dielectric constant
λ	m	Tunnel parameter
λ_c	m	Tunneling depth for capturing
λ_e	m	Tunneling depth for emission
λ_0		Slope of the RTS scaling trend
μ	$m^2 V^{-1} s^{-1}$	Mobility
μ_0	$m^2 V^{-1} s^{-1}$	Low field mobility
σ	Ω^{-1}	Channel conductivity
σ_c	m^2	Effective capture cross section
σ_h	m^2	Capture cross section for holes
σ_{IR}	A	Standard deviation in the read current of an RRAM cell
σ_n	m^2	Capture cross section for electrons
σ_{no}	m^2	Capture cross section at the interface for electrons
σ_{var}	A	Standard deviation of a variable
σ_t	Ω^{-1}	Conductivity around a charged trap
σ_0	m^2	Temperature independent cross-section pre-factor
τ_c	s	Capture time constant
$\tau_{cA\bar{B}}$	s	Capture time constant of trap A with trap **B** empty
τ_{cAB}	s	Capture time constant of trap A with trap B occupied
τ_e	s	Emission time constant
τ_{tc}	s	Capture time via a tunnel process
τ_{te}	s	Emission time via a tunnel process

τ_1	s	Average time constant
ϕ_e	V	Electron affinity difference between Si and SiO_2
ϕ	V	Trap potential
Ψ_s	V	Band bending at the semiconductor surface
$\delta\Psi_s^T$	V	Surface potential shift near a trap
ϖ	s^{-1}	Phonon frequency

List of abbreviations

AC	Alternating current
AHT	As-grown hole trap
BD	Breakdown
BOX	Buried oxide
BTI	Bias-temperature instability
CB	Conduction band
CC	Configuration coordinate
CCD	Charge-coupled device
CESL	Contact etch stop layer
CET	Capacitance equivalent thickness
CEMT	Capture/emission time
CF	Conducting filament
CMOS	Complementary metal–oxide–semiconductor
CNT	Carbon nanotube
DC	Direct current
DCR	Dark count rate
DF	Duty factor
DFT	Density functional theory
DLTS	Deep level transient spectroscopy
DRAM	Dynamic random access memory
EDT	Edge direct tunneling
ENOT	Effective number of traps
EOT	Equivalent oxide thickness
ET	Extremely thin
FD	Fully depleted
FDSOI	Fully depleted silicon-on-insulator
FET	Field-effect transistor
FF	FinFET
FG	Floating gate
FGA	Forming gas annealing
FHMM	Factorial hidden Markov model
FUSI	Fully silicided
GAA	Gate-all-around
GD	Generated defect
GIDL	Gate-induced drain leakage
GR	Generation–recombination
HBD	Hard breakdown
HC	Hot carrier
HCI	Hot carrier injection
HRS	High-resistance state
HTFGA	High-temperature forming gas annealing
ICP	Inductively coupled plasma
IL	Interfacial layer
ITRS	International Technology Roadmap for Semiconductors
JL	Junctionless

LER	Line edge roughness
LFN	Low frequency noise
LOCOS	Local oxidation of silicon
LRS	Low resistance state
MC	Monte Carlo
MGG	Metal gate granularity
MOSFET	Metal–oxide–semiconductor field-effect transistor
MPTAT	Multi-phonon trap-assisted tunneling
NBTI	Negative bias-temperature instability
NVM	Non-volatile memory
NW	Nanowire
NWFET	Nanowire FET
PBTI	Positive bias-temperature instability
PCM	Phase change memory
PP	Percolation path
PSD	Power spectral density
PV	Process variability
P/E	Program/erase
RDF	Random doping fluctuation
ReRAM	Resistive random access memory
RMG	Replacement metal gate
RTA	Rapid thermal annealing
RTN	Random telegraph noise
RTS	Random telegraph signal
RW	Random walk
SBD	Soft breakdown
S/D	Source/drain
SDRAM	Synchronous dynamic random access memory
SET	Single-electron transistor
SHR	Shockley–Read–Hall
SILC	Stress-induced leakage current
SNWT	Silicon nanowire transistor
SOI	Silicon-on-insulator
SON	Silicon-on-nothing
SPAD	Single-photon avalanche photodiode
SPICE	Simulation program with integrated circuit emphasis
SRAM	Static random access memory
SS	Subthreshold swing
STI	Shallow trench isolation
STID	Single-trap-induced degradation
STM	Scanning tunneling microscopy
TAT	Trap-assisted tunneling
TCAD	Technology computer-aided design
TDDS	Time-dependent defect spectroscopy
TEG	Test element group
TFET	Tunnel FET
TLF	Two-level fluctuation
TLP	Time-lag plot

TSS	Two-state system
VB	Valence band
VJL	Variable junction leakage
WF	Work function
WID	Within die
WKB	Wentzel–Kramers–Brillouin
2DEG	Two-dimensional electron gas

Chapter 1

Introduction

Since the first report by Ralls *et al* in 1984 [1], random telegraph signals (RTSs) have caught the attention of the device-physics community and, more recently, also device technologists and circuit designers. Obviously, scaling of the metal–oxide–semiconductor field-effect transistor (MOSFET) feature size (length L, width W) has reduced the number of active traps at the Si/SiO_2 interface in such a way that only one or a few defects affect the charge transport in a nanoscale transistor. In extreme cases, the current fluctuations induced by a single (giant) trap may amount to nearly 100% of the static current, making the device useless as a logic element.

In its simplest form, RTS consists of the temporal switching of the drain current I_D (conductance $G_D = I_D/V_{DS}$ in linear operation) or the drain voltage V_{DS} (channel resistance $R_D = V_{DS}/I_D$) between two levels, as illustrated by figure 1.1 [2]. The random switching between the high and the low current level is characterized by the time spent in the high state t_h and in the low state t_l, and the switching amplitude ΔI_D. The discrete nature of the RTS fluctuations and the characteristic order of magnitude have been interpreted in terms of trapping of a single electron (nMOSFET) or hole (pMOSFET) by a defect at the Si/SiO_2 interface or in the oxide at tunneling distance from the inversion layer (1–2 nm typically). The latter may also be termed border traps, which are able to communicate with the inversion layer on a millisecond to >second time scale. As such, it has been suggested that RTS is the fundamental component of $1/f$ noise in MOSFETs, which is then composed of the summation of different RTSs, with various trapping/detrapping time constants [3].

It is the aim of this book to summarize the knowledge related to the physics and phenomenology of RTS, as interest in this phenomenon has exploded in recent years due to the awareness that RTS may be of concern in the scaling of deep submicron memories (flash memory, static random access memory (SRAM), dynamic random access memory (DRAM)) and in complementary metal–oxide–semiconductor (CMOS) imagers. In addition, RTS contributes to the dynamic process variability

doi:10.1088/978-0-7503-1272-1ch1

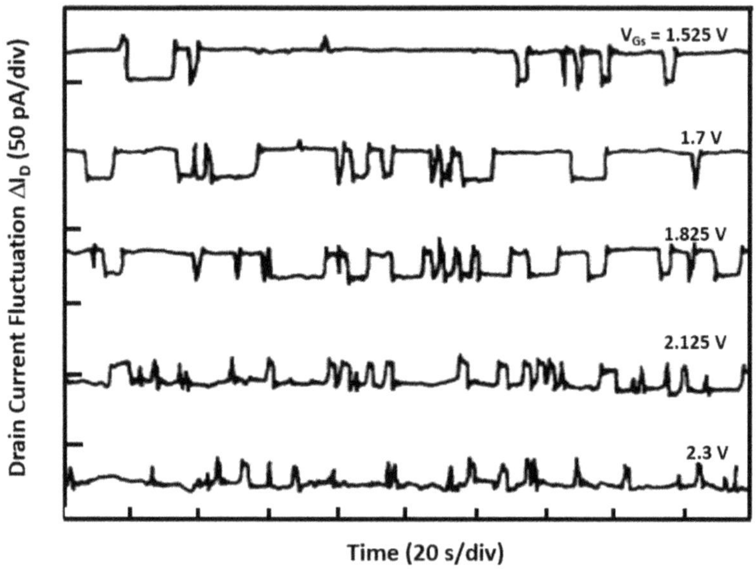

Figure 1.1. RTSs in a small MOSFET measured at the indicated gate voltages. The active device area is 0.4 μm^2, $V_{DS} = 4$ mV, $T = 293$ K. (Reproduced with permission from Kirton *et al* [2]. Copyright IOP Publishing.)

(PV) of CMOS logic devices and circuits and may overtake random doping fluctuations (RDFs) as its dominant source, particularly for extremely scaled technologies (semiconductor nanowires (NWs), carbon nanotube (CNT) FETs, etc). Finally, RTS also appears to be an important factor in the understanding of recent reliability issues related to negative and positive bias temperature instability (NBTI and PBTI) in state-of-the-art CMOS technologies.

In chapter 2, the phenomenology of RTS in small-area MOSFETs will be described, whereby the dependence of t_l and t_h on bias conditions (gate voltage V_{GS}, drain voltage V_{DS} and substrate voltage V_{BS}) will be highlighted. Likewise, the behavior of the RTS amplitude for different device sizes and operation conditions will be presented. Distinction will be made between RTSs in the channel and in the gate current. The latter become important for a scaled gate oxide thickness t_{ox}, where the probability for direct tunneling of carriers is no longer negligible, leading to a measurable gate current I_G. It is also shown that complex, multi-level RTSs may exist in I_G, which belong to a special class of metastable or closely spaced, interacting defects, e.g. a defect cluster. In addition, anomalous two-level RTSs exist, having huge amplitudes, close to 100% of the direct current (DC). Upon scaling the transistor dimensions, the off-state leakage current is no longer negligible and can be composed of gate-induced drain leakage (GIDL) and drain–substrate leakage current. As will be seen, RTS occurs in these leakage currents as well and may be caused by specific, interface-state related mechanisms.

Chapter 3 describes in more detail the physics-based modeling of RTS in state-of-the-art and nanometer devices, including inversion layer quantization, Coulomb blockade and image charge effects. The observed wide scatter in the RTS amplitude can be explained by considering the stochastic nature of the surface potential at the Si/SiO_2 (or high-κ) interface, induced by the random distribution of dopants in the substrate and fixed charges in the gate dielectric. This implies a mutual influence between spatial RDFs and temporal trapped-charge fluctuations. The spread in the RTS amplitude follows from the random placement of a trap with respect to the critical current paths (or percolation filament). A procedure will be outlined for extracting the trap-level parameters from an RTS. In addition to its activation energy E_T and capture cross section for electrons σ_n (nMOSFET) or holes σ_p (pMOSFETs), it is possible to derive the trap depth (x_T) and lateral position along the channel (y_T). Such analysis can, in principle, be extended to gate-current RTS. One of the drawbacks of RTS measurements is that they are lengthy and they require a large set of data under different conditions to yield meaningful results. Therefore, as is discussed in chapter 3, several novel approaches have been proposed which speed up the analysis without compromising too much on accuracy. Finally, in recent years, there has been significant progress in the numerical modeling of RTS based on an atomistic approach. This has led to a clear understanding of the wide spread in RTS amplitudes that is observed experimentally. In combination with *ab initio* density functional theory (DFT) calculations, this opens the path to the most important question: *what is the physical origin of the traps responsible for RTS*?

In chapters 4 and 5 technological (scaling) and operational (reliability) factors that influence the occurrence of RTS will be discussed. It is clear that the quality of the gate oxide will have a strong impact on the number of active traps. This applies in general to all processing steps which can cause the gate oxide quality to deteriorate and, hence, create process-induced RTSs (plasma etching, rapid thermal annealing (RTA), ion implantation, nitridation, a high-κ gate stack, etc). An important factor appears to be the lateral device isolation (local oxidation of silicon (LOCOS), shallow trench isolation (STI)), which may induce a higher number of traps, located at the high electric field edges of the devices. Conversely, thermal curing and passivation steps assist in removing RTSs from the devices. In addition, stressing of the transistors, employing either uniform (Fowler–Nordheim (FN), NBTI or PBTI, irradiation) or non-uniform (hot-carrier (HC), degradation) stressing conditions, can create additional RTSs both in the channel and the gate current. As will be shown, the nature of the stress-induced RTSs is generally different from the process-induced ones. At the same time, the step-wise relaxation transients of small-area nFETs and pFETs after BTI stress strongly suggest that they are related to the discharging of individual traps in the gate dielectric, similarly to RTS. By analogy for long channel transistors, the BTI relaxation following a log(time) dependence shares the same origin as the $1/f$ dependence of the low-frequency noise (LFN) power spectral density (PSD), namely detrapping from a large ensemble of gate oxide traps with a large distribution of emission time constants. Finally, RTS in exploratory and beyond-silicon devices is discussed at the end of chapter 4.

In chapter 6 the impact of RTSs on scaled flash, SRAM and DRAM memories, on non-volatile memories (NVM) and on imagers (CMOS imagers, charge-coupled devices (CCDs)) will be treated. As will be shown, in order to evaluate the effect of RTS on circuit operation, a statistical evaluation is essential, relying on the measurement of a large number of devices. After all, RTS is still quite a rare event, even in state-of-the-art technologies, occurring in less than 1% of transistors. As will be seen, there is ample evidence for the existence of extreme or anomalously large RTSs which threaten the reliable operation and functionality of memories, and may become a bigger issue than the PV associated with RDF around the 14 nm node or below. Some general conclusions will be given in chapter 7.

Lastly, it should be noted that the early work on single-device RTSs has been reviewed on several occasions. More details can be found in [2–8]. Throughout the book RTS is illustrated and discussed in a variety of semiconductor devices, and it is assumed that the reader is familiar with the basic operation and fabrication aspects of these devices. For further details see the existing books and review articles covering devices such as CMOS process technology [9, 10], silicon-on-insulator (SOI) technology [11], FinFETs (FFs) [12–14], DRAM [15], NVM [16, 17], phase change memory (PCM) and resistive random access memory (ReRAM) [18], and CNTs [19–21].

References

[1] Ralls K S, Skocpol W J, Jackel L D, Howard R E, Fetter L A, Epworth R W and Tennant D M 1984 Discrete resistance switching in submicrometer inversion layers: individual interface traps and low-frequency (1/*f*?) noise *Phys. Rev. Lett.* **52** 228–31

[2] Kirton M J, Uren M J, Collins S, Schulz M, Karmann A and Scheffer K 1989 Individual defects at the Si:SiO_2 interface *Semicond. Sci. Technol.* **4** 1116–26

[3] Kirton M J and Uren M J 1989 Noise in solid-state microstructures: a new perspective on individual defects, interface states and low-frequency (1/*f*) noise *Adv. Phys.* **38** 367–468

[4] Restle P and Gnudi A 1990 Internal probing of submicron FETs and photoemission using individual oxide traps *IBM J. Res. Dev.* **34** 227–42

[5] Mueller H H and Schulz M 1995 Review—individual interface traps at the Si–SiO_2 interface *J. Mater. Sci.: Mater. Electron.* **6** 65–74

[6] Ghibaudo G 2000 Critical MOSFETs operation for low voltage/low power ICs: ideal characteristics, parameter extraction, electrical noise and RTS fluctuations *Microelectron. Eng.* **39** 31–57

[7] Ghibaudo G and Boutchacha T 2002 Electrical noise and RTS fluctuations in advanced CMOS devices *Microelectron. Reliab.* **42** 573–82

[8] Simoen E and Claeys C 2002 Random telegraph signal: a local probe for single point defect studies in solid-state devices *Mater. Sci. Eng.* B **91-92** 136–43

[9] Plummer J D, Deal M D and Griffin P B 2000 *Silicon VLSI Technology* (Upper Saddle River, NJ: Prentice Hall)

[10] Wolf S and Tauber R N 2000 *Silicon Processing for the VLSI Era: Process Technology* (Sunset Beach, CA: Lattice Press)

[11] Kononchuk O and Nguyen B Y 2014 *Silicon-on-Insulator Technology* (Amsterdam: Elsevier)

[12] Colinge J-P 2007 *FinFET and other Multi-gate Transistors* (Berlin: Springer)

[13] Horiguchi N 2011 FinFETs and their future *Semiconductor-on-Insulator Materials for Nanoelectronics Applications* ed A Nazarov *et al* (Berlin: Springer) pp 141–53

[14] Collaert N 2013 *CMOS Nanoelectronics: Innovative Devices, Architectures and Applications* (Singapore: Pan Stanford)

[15] Schroeder H and Kingon A 2003 High-permittivity materials for DRAMs *Nanoelectronics and Information Theology—Advanced Electronic Materials and Novel Devices* ed R Waser (New York: Wiley) pp 541–63

[16] Groeseneken G, Maes H E, Van Houdt J and Witters J S 1997 Basics of nonvolatile semiconductor memory devices *Nonvolatile Semiconductor Memory Technology, a Comprehensive Guide to Understanding and Using NVM Devices* ed D B William and E B Joe (Piscataway, NJ: IEEE) pp 1–88

[17] Van Houdt J, Degraeve R, Groeseneken G and Maes H E 2007 Nonvolatile memory technologies with emphasis on flash, a comprehensive guide to understanding and using NVM devices *Physics of Flash Memories,* ed J Brewer and M Gill (Piscataway, NJ: IEEE) pp 129–77

[18] Wouters D J, Waiser R and Wuttig M 2015 Phase-change and redox-based resistive switching memories *Proc. IEEE* **103** 1274–88

[19] Derycker V, Filorama A and Bourgoin J-P 2009 Carbon nanotube electronics *Nano-CMOS Era—From Ultimate CMOS Scaling to Beyond CMOS Devices* ed S Deleonibus (Singapore: Pan Stanford) pp 333–64

[20] Che Y, Chen H, Gui H, Liu J, Liu B and Zhou C 2014 Review of carbon nanotube nanoelectronics and macroelectronics *Semicond. Sci. Technol.* **29** 073001

[21] Chen Z 2014 Nanotubes for nanoelectronics *Encyclopedia of Nanoscience and Nanotechnology* ed H S Nalwa (Valencia, CA: American Scientific) pp 919–42

IOP Publishing

Random Telegraph Signals in Semiconductor Devices

Eddy Simoen and Cor Claeys

Chapter 2

Random telegraph signal phenomenology

This chapter summarizes the experimental behavior of a typical individual RTS in single, small-area MOSFETs. A simple RTS is characterized by three parameters: its up (or high) and down (or low) time and the switching amplitude. The gate voltage dependence of a drain current (I_D) RTS allows one, in principle, to identify whether t_h and t_l correspond to the carrier capture or the emission time (or vice versa), and reveals if the trap has a donor or acceptor character. As is shown, for most cases at or near room temperature the time constants are thermally activated, giving rise to a straight Arrhenius plot versus $1/T$ (T is the temperature). From this, the energy level E_T and the barrier for carrier capture (E_B) can be derived. At low(er) temperatures, where thermal activation is less likely, RTSs may be found, which are characterized by tunneling transitions with negligible activation energy. From the variation of the average emission (τ_e) and capture (τ_c) time constants on the gate and drain bias, the depth and lateral trap position can be extracted. While standard Shockley–Read–Hall (SRH) theory applies in many cases (particularly for a trap in strong inversion), the Coulomb charging effect necessitates an improved description of the capture and emission kinetics, in particular for thinner gate oxides and in weak inversion (around the threshold voltage V_T). At the same time, the RTS amplitude is readily evaluated, while the large spread in values found in practice indicates that it cannot be simply modeled as resulting from pure number fluctuations, i.e. an amplitude given by $1/N$ with N the total number of carriers in the channel. This highlights the importance of the mobility fluctuation (or scattering) contribution $\Delta\mu/\mu$. However, as will be shown, other factors may also affect ΔI_D, resulting in anomalously large amplitudes. RTS has also been studied with the gate biased in accumulation, thereby analyzing the GIDL current. This enables the study of different gate oxide traps compared to normal inversion operation. In the next section, the parameters of RTSs in the gate current will be outlined, demonstrating that they behave differently from the channel RTSs, most probably because they belong to a different kind of trap. It will also be shown that RTS fluctuations in the gate current can be associated with the channel or the gate electrode, depending on the polarity of the time constant ratio. Next, the occurrence of RTS in the drain–junction leakage current will be discussed, which is potentially of great interest for the retention time in DRAMs. In the final section, the occurrence of complex RTSs, generated by multiple traps, will be outlined. While in many cases it is assumed that the traps are independent of each other, it will be shown that mutual interactions cause specific phenomena, which should be accounted for in device and circuit operation.

2.1 RTS time constants

The times the current stays in the up or down level are stochastically distributed and for an individual, non-interacting trap generally follow an exponential distribution (figure 2.1 for the down time), represented by

$$P_1(t) = \frac{1}{\tau_1}\exp\left(-\frac{t}{\tau_1}\right) \tag{2.1}$$

with τ_1 the average time constant and $P_1(t)\mathrm{d}t$ the probability that the high state 1 will not make a transition for time t, then will make one in the interval t to $t + \mathrm{d}t$.

2.1.1 The SRH framework

Assuming that the trap is at the Si/SiO_2 interface (or in its close vicinity), one can in the first instance describe the carrier capture and emission by a trap in the frame of

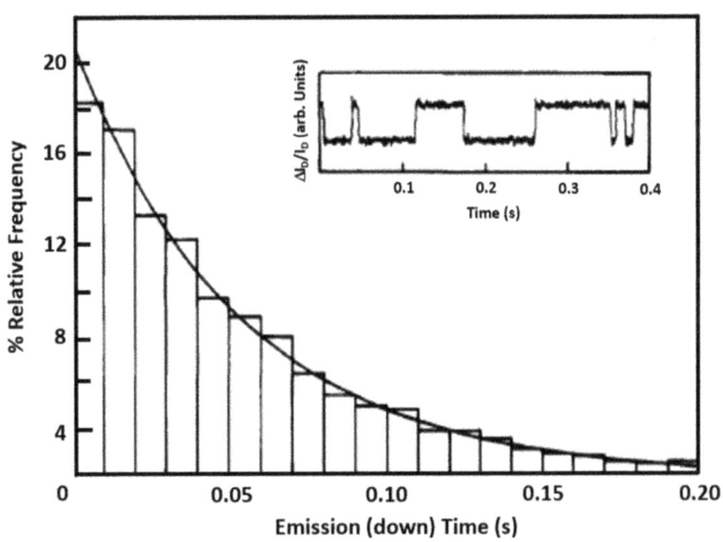

Figure 2.1. Distribution of 4425 emission times at 95 K and $V_{GS} = 1.15$ V, showing that the time is distributed exponentially. $\tau_e = 0.0528$ s, standard deviation = 0.505 s. The inset shows a portion of the I_D–t characteristic; the down time corresponds to emission. (Reproduced with permission after Kirton *et al* [1]. Copyright 1989 IOP Publishing.)

the SRH theory [2, 3]. In that case, the average capture time constant for an electron in the inversion layer is given by

$$\tau_c = \frac{1}{n\sigma_n v_{thn}} \tag{2.2}$$

with n the volume concentration (in cm^{-3}) and v_{thn} the thermal velocity for electrons. In principle, n is proportional to the drain current I_D [3]; it is tacitly assumed that *inversion layer quantization can be neglected.* More accurate treatments, including inversion layer quantization will be given later. The use of v_{thn} in equation (2.2) considers an electron gas in equilibrium with the lattice, *with negligible carrier heating* in the lateral electrical field V_{DS}/L (linear operation). This is warranted as long as V_{DS}/L is kept small (i.e. on the order of kT) and for not too low temperatures.

From the principle of detailed balance (equal numbers of up and down jumps) it follows that [2]:

$$\frac{\tau_c}{\tau_e} = g \exp\left(\frac{E_T - E_F}{kT}\right) = \frac{1 - f_T}{f_T} \tag{2.3}$$

with E_F the surface Fermi level and k Boltzmann's constant. The ratio of the capture and emission time constant is also called the mark space ratio, where f_T is the trap occupancy by an electron, given by the Fermi–Dirac function. The trap degeneracy g is usually not known and is assumed to be equal to 1.

Taking into account the fact that $n = N_C \exp[(E_C - E_F)/kT]$, with N_C and E_C the density of states and the bottom of the 3D conduction band (CB), respectively, readily yields [1]

$$\tau_e = \frac{1}{N_C v_{thn} \sigma_n} \exp\left(\frac{E_C - E_T}{kT}\right). \tag{2.4}$$

Symmetric relations hold for a p-channel device, replacing σ_n by σ_p, v_{thn} by v_{thp} and $E_C - E_T$ by $E_T - E_V$ (E_V is the maximum of the valence band).

The first practical question is to identify t_l and t_h with either the capture or the emission time. In the frame of SRH theory, one expects that τ_c decreases with I_D and, hence, with V_{GS}. In other words, the electron occupancy should increase with V_{GS} or I_D (nMOSFET). This is illustrated in figure 2.2 [4] for an RTS in a 0.15 μm^2 nMOSFET with a 4 nm SiO_2 gate dielectric in linear operation. It is also clear that the device and RTS characteristics are symmetric, independent of forward (F) or reverse (R) operation. The emission time, on the other hand, increases with V_{GS}. This is the typical behavior for acceptor type RTSs in nMOSFETs, changing charge state from more positive to more negative by electron capture. The low current state corresponds to the trap occupied by an electron, while in the high state the trap is empty. From equation (2.3) one derives that the mark space ratio decreases exponentially with $-E_F$ or with $-V_{GS}$, which is illustrated in figure 2.3 [4]. Also shown is the drain voltage dependence, peaking around 0.1 V. As will be seen next, this allows one to extract the lateral position of the trap along the Si/SiO_2 interface [4].

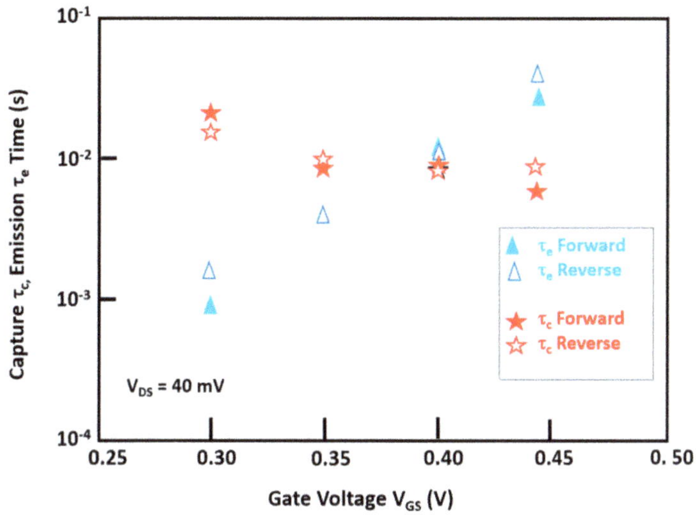

Figure 2.2. Mean capture and emission times as a function of V_{GS} for $V_{DS} = 40$ mV. Reverse is with the source and drain switched. (Reprinted with permission from Amarasinghe *et al* [4]. Copyright 2000 Elsevier.)

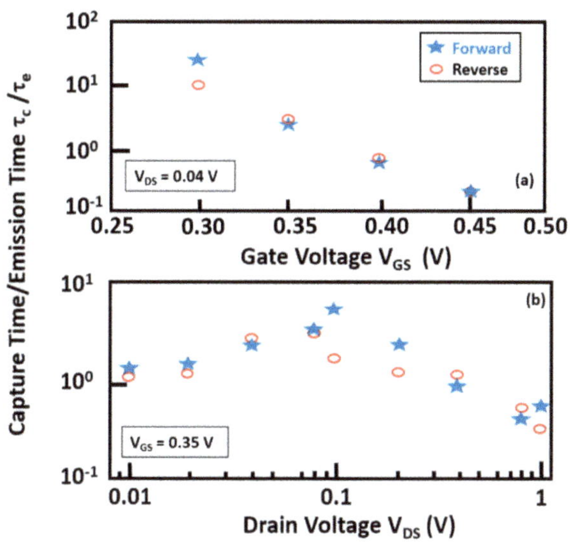

Figure 2.3. *(a)* τ_c/τ_e versus V_{GS} for $V_{DS} = 40$ mV. The slope is proportional to x_T, the depth position of the trap in the oxide. *(b)* τ_e/τ_c versus V_{DS} for $V_{GS} = 0.35$ V. Reverse operation is with source and drain switched. (Reprinted with permission from Amarasinghe *et al* [4]. Copyright 2000 Elsevier.)

Finally, from the weak substrate bias dependence noticed in figure 2.4, it has been concluded that the vertical electric field in the oxide has only a modest impact on the time constants [5]. The field dependence should correspond to a decrease with $-q(V_{GS} - V_T)x_T/t_{ox}$ of the activation energy in strong inversion [4] and is typically a few meV [5].

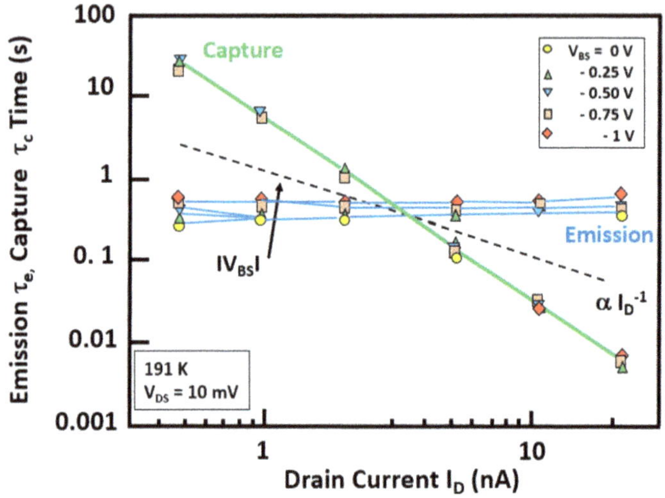

Figure 2.4. Mean emission and capture times as a function of the drain current I_D at $V_{DS} = 10$ mV, with the substrate bias V_{BS} as a parameter. The dotted line indicates the capture SRH law proportional to I_D^{-1}. (Reproduced with permission from Mueller *et al* [5]. Copyright 1994 AIP Publishing.)

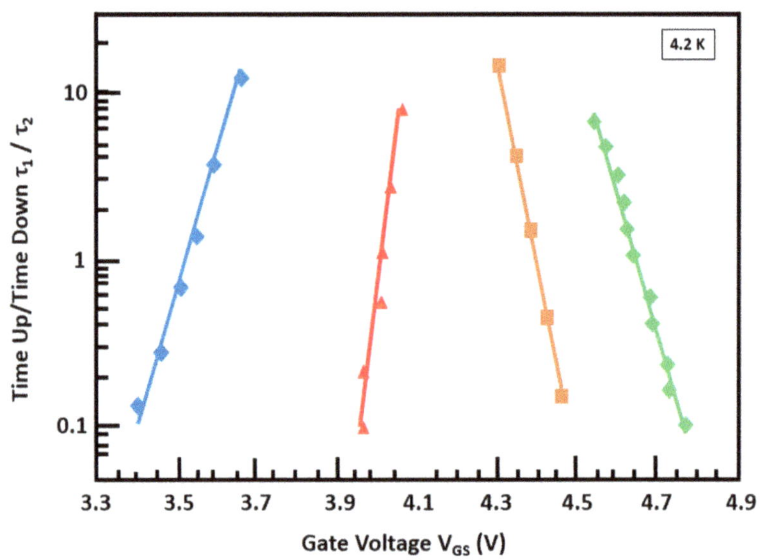

Figure 2.5. Gate-voltage dependence of the up/down ratio of four two-level RTSs in unstressed nMOSFETs at 4.2 K. (Reprinted with permission from Cobden and Uren [6]. Copyright 1993 Elsevier.)

At low(er) temperatures, one can find RTSs exhibiting a different polarity [2, 6]. In other words τ_c/τ_e increases with V_{GS} in an nMOSFET, as illustrated by figure 2.5 [6]. This implies that equation (2.3) can be generalized to [6]

$$\frac{\tau_c}{\tau_e} = \exp\left(\frac{qs\eta(V_{GS} - V_{GS0})}{kT}\right) \tag{2.5}$$

with q the elementary charge, s the polarity (+1 or −1) and $\eta = x_T/t_{ox}$. V_{GS0} is the gate voltage for which $\tau_c = \tau_e$. These traps with a positive polarity usually have a smaller amplitude [2] and are believed to be associated with attractive traps [7], i.e. a donor for n-channel transistors, switching between + (empty—capture) and 0 (full—emission). It has been argued that in this case the amplitude is not due to charge exchange (number fluctuations), but rather to energy exchange (scattering, mobility) [6, 7].

As mentioned before, at low temperatures and sufficiently high V_{DS}, carrier heating can take place in the lateral electric field in the channel [8, 9]. This will affect the carrier thermal velocity in equation (2.2), according to [8, 9],

$$v_n = \sqrt{v_{thn}^2 \frac{T_e}{T} + v_y^2} \tag{2.6}$$

with T_e the electron temperature (T the lattice temperature) and v_y the drift velocity that can be approximated by $qV_{DS}y_T/L$ [10]. Consequently, the capture time will be decreased and is described by [8]

$$\tau_c = \frac{v_{thn}\tau_{c0}}{v_n} e^{\frac{\Delta E_c}{k}(1/T_e - 1/T)} \tag{2.7}$$

with τ_{c0} the capture time at low V_{DS}. The non-equilibrium electron temperature T_e can be estimated from the measured field-dependent mobility μ as [8]:

$$T_e = \frac{\mu_0}{\mu} T \tag{2.8}$$

with μ_0 the low-field mobility. Based on this model, good agreement with experimental behavior has been found [8], as can be noted in figure 2.6.

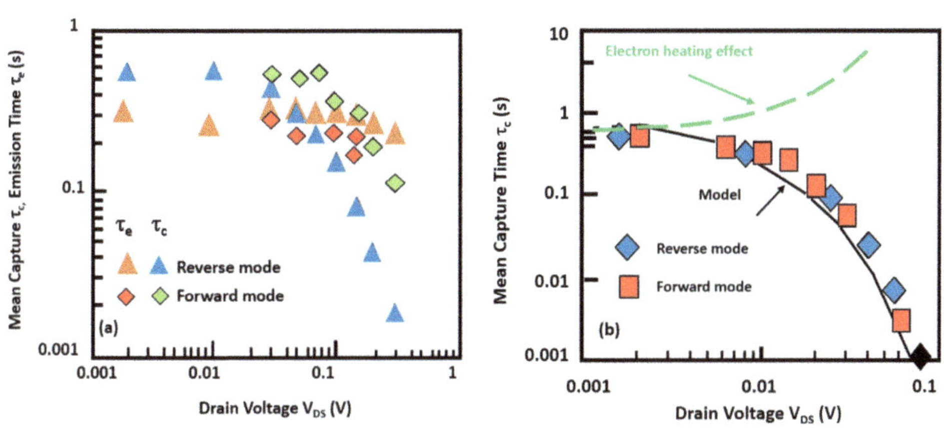

Figure 2.6. (*a*) Mean capture and emission times versus V_{DS} with V_{GS} in weak inversion for forward (open symbols) and reverse modes (filled symbols). (*b*) Scaled and theoretically predicted mean capture time. The dotted line gives the effect of electron heating. $W \times L = 2 \times 0.1$ μm^2; $t_{ox} = 6$ nm, $V_{GS} = 0.34$ V, $T = 86$ K. (Reproduced with permission from Shi *et al* [8]. Copyright 1994 IOP Publishing.)

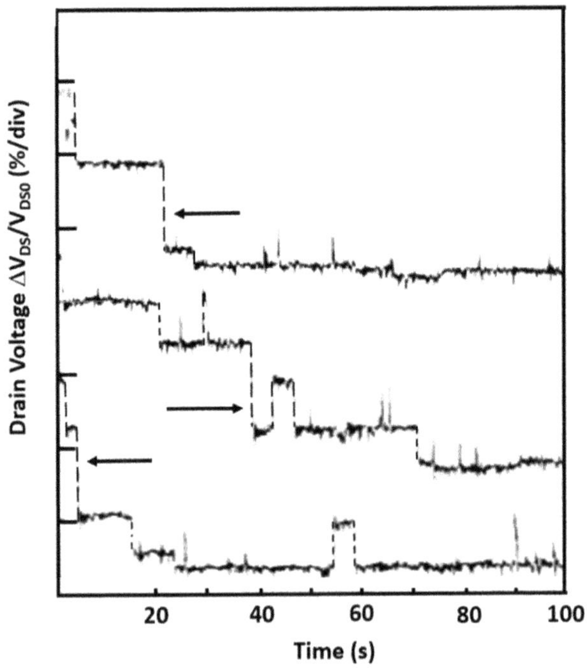

Figure 2.7. Quantized transients of electron re-emission from individual traps in an n-channel MOSFET. The voltage drop across the channel is monitored versus time for a constant current, as indicated. Sequential recordings of transients under the same conditions are vertically displaced to avoid superposition. The arrows mark the occurrence of step height in the sequential transients. Device dimensions: 1.2 μm × 1.0 μm. V_{GS} = 0.66 V; V_{DS} = 10 mV; I_D = 20 nA; T = 295 K. (Reproduced with permission from Kirton *et al* [1]. Copyright 1989 IOP Publishing.)

In addition to the study of RTS in the time (or frequency) domain at steady-state, the use of non-steady-state conductance deep level transient spectroscopy (DLTS) on small-area MOSFETs has also been reported [1, 11, 12]. In this case, the voltage transient at a constant drain current is monitored following a trap-filling gate-voltage pulse from inversion into accumulation. This gives rise to quantized transients, as in figure 2.7, where each discrete step corresponds to the re-emission of an electron (nMOSFET) from a specific trap site. The number of downward steps corresponds to the number of active interface (or near-interface oxide) traps, so that they can be counted for each operation condition (V_{GS}, T). Repeating the transient measurement yields the same step height at a different emission time t_e, so that in this way it is possible to derive the distribution of emission time constants at a certain temperature T. The average τ_e is then determined at different temperatures to yield an Arrhenius plot (see the following section).

Quite recently, the DLTS principle has been extended for the development of so-called time-dependent defect spectroscopy (TDDS) [13, 14], which has been proposed as a technique to study relaxation transients after a bias-temperature trap-filling gate-voltage stress. More details on reliability will be given in chapter 5.

2.1.2 Trap energy, capture barrier and location from the SRH approach

In many cases, the average capture time appears to be thermally activated [1–3, 15]. This means that equation (2.2) for electron capture *from a non-quantized inversion layer*, assumed in this section, can be rewritten as [3]

$$\tau_c = \frac{\exp\left(\frac{E_B}{kT}\right)}{n\sigma_0 v_{thn}} \tag{2.9}$$

with E_B the energy barrier for electron capture and σ_0 the temperature-independent cross-section pre-factor. This activation energy can be interpreted in terms of a strong lattice or structural relaxation taking place in the amorphous gate oxide by emission of multiple phonons upon carrier capture, dissipating part of the excess energy of the trapped carrier. This non-elastic, multi-phonon-assisted tunneling takes into account electron–phonon interactions, which give rise to a rearrangement of the atoms in the defect structure caused by the electrostatic coupling between the nuclei and the trapping electron [16]. Such a structural rearrangement, which requires the atoms to leave their equilibrium positions, is associated with the system overcoming an energy barrier, which retards the electron trapping process and is given by [16]

$$E_B = \frac{(E_{relax} - E_0)^2}{4E_{relax}} \tag{2.10}$$

E_{relax} and E_0 are defined in the configuration coordinate (CC) diagram of figure 2.8, showing the elastic (vibrational) plus electronic energy of the system versus a single

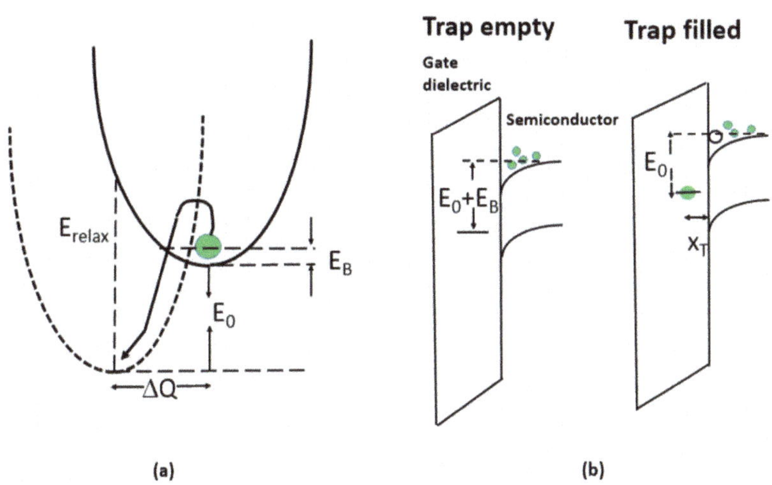

Figure 2.8. Total energy (*a*) and single-electron (*b*) diagrams (Q is the generalized coordinate of the atomic displacements in the system). The 'initial' (solid line) and 'final' (dashed line) states correspond to the system with an empty and filled trap, respectively. E_{relax} is the energy corresponding to displacements of the lattice atoms, ΔQ, caused by electron trapping. E_0 is the energy lost by the electronic subsystem after a trapping event. (*a*) indicates that electron trapping is associated with the system transitioning over the barrier E_B. (After Veksler *et al* [16].)

normal coordinate Q [16]. The capture barrier height depends on the trap energy and on the gate bias. The relaxation energy E_{relax} is a measure of the displacement of the lattice atoms around the defect caused by the trapped charge. Substituting equation (2.9) in equation (2.4) thus shows that the activation energy corresponding to the average emission constant is now determined by the sum of $\Delta E_{CT} = E_C - E_T$ and E_B.

Furthermore, ΔE_{CT} should be regarded as a Gibbs free energy change, which is composed of a change in enthalpy ΔH_{CT} and entropy $T\Delta S$ [3]. Substituting this in equation (2.4) yields [3]

$$\tau_e = \frac{\exp\left(-\frac{\Delta S}{k}\right)}{N_C v_{thn} \sigma_{n0}} \exp\left(\frac{\Delta H_{CT} + E_B}{kT}\right) \tag{2.11}$$

.

Using an Arrhenius analysis of the emission and capture time constants, the activation enthalpy ΔH_{CT}, the capture barrier E_B and the entropy change ΔS have been determined for a number of RTSs in n-channel [3] and p-channel [17] transistors. Generally, the barrier for capture (several hundreds of meV) is higher than the activation enthalpy (~100 meV). It was also found that the entropy change for the trapping of a hole is similar to that for the trapping of an electron [17], in the range of 5 to 10 k. The sign of the change is different for electrons than for holes, indicating that different kinds of traps are responsible for electron and hole trapping at the Si/SiO_2 interface. The entropy change for electrons implies that the energy level shifts towards the CB with increasing temperature, while the opposite holds for hole traps in pMOSFETs, i.e. the hole trapping level shifts towards the valence band with increasing temperature. Entropy changes in the range of 5 to 10 k are not unrealistic for transitions involving significant lattice relaxation in a material with strong ionicity (an oxide) [17]. From this different entropy change, it was concluded that the slow states responsible for RTS are in fact connected to the silicon band structure, whereby the density increases towards the band edges, similarly to the true interface states (a U-shaped profile).

It was also concluded that the range of observed E_B values is consistent with capture into defects in an amorphous material, with its consequent continuous distribution of trap environments [1]. At the same time, the σ_0 pre-factors appear to be correlated with the E_B values, with higher barriers (~0.5–0.6 eV) corresponding to higher σ_0 values (~a few 10^{-15} cm^2). It was argued that this is probably coincidental and related to the limited time constant range accessible by the steady-state RTS measurements. Indeed, from equation (2.9) it can be derived that, for a fixed capture time, a high σ_0 value will be compensated by a high E_B and vice versa. Giving the limited maximum observation time (typically tens of seconds), the window of accessible gate oxide traps is restricted to the condition where the capture time is of the same order as the emission time.

Considering the band diagram of figure 2.9 for an nMOSFET and linear operation ($V_{DS} \ll (V_{GS} - V_T)$), equation (2.4) can be transformed into [4]

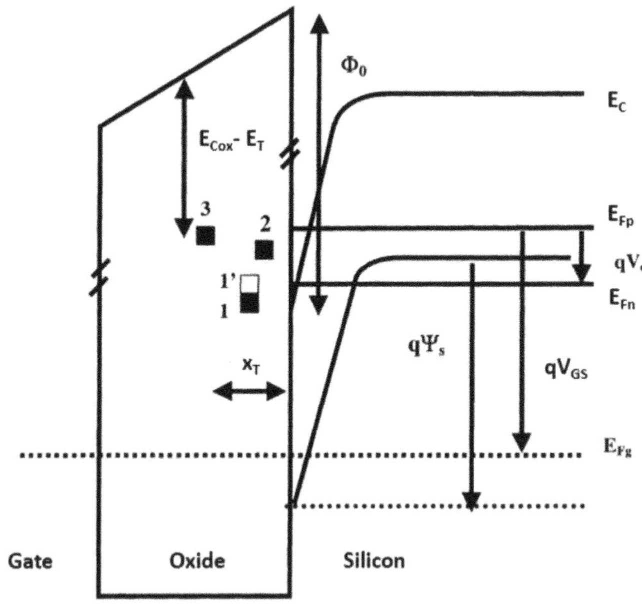

Figure 2.9. The energy band diagram in the channel of an nMOSFET, showing the depth and energy of different oxide traps. Here, V_c is the channel voltage equal to $(E_{Fp} - E_{Fn})/q$, the respective quasi-Fermi levels for holes and electrons. (Reproduced with permission from Amarasinghe and Çelik-Butler [18]. Copyright 2000 Elsevier.)

$$\ln\frac{\tau_c}{\tau_e} = -\frac{1}{kT}\Big[(E_{Cox} - E_T) - (E_C - E_F) - \phi_e + q\Psi_s + qx_T/t_{ox}(V_{GS} - V_{FB} - \Psi_s)\Big]. \tag{2.12}$$

E_{Cox} is the CB minimum in the oxide, ϕ_e is the difference in the electron affinities, Ψ_s is the amount of band bending in the semiconductor and V_{FB} is the flat-band voltage. From equation (2.12) it then immediately follows that the trap depth is given by

$$x_T = -\frac{kTt_{ox}}{q}\frac{\mathrm{d}\left(\ln\frac{\tau_c}{\tau_e}\right)}{\mathrm{d}V_{GS}} \tag{2.13}$$

This is valid for a MOSFET *in strong inversion operation* ($V_{GS} > V_T$), so that the quasi-Fermi level for electrons (holes) is approximately pinned to the conduction (valence) band and the term $\partial\Psi_s/\partial V_{GS}$ can be neglected. In other words, the trap position is given by the shift of the activation energy with gate bias. It should be remarked that equation (2.12) *is only valid for traps obeying the SRH theory*, i.e. with a *negligible Coulomb blockade effect*. In practice, this means acceptor traps in strong inversion [4].

In case of a Coulomb blockade present in a repulsive trap (acceptor for nMOSFET), it is better to measure the RTS at the same current I_D with a substrate bias V_{BS} and at 0 V [19]. This yields

$$x_T = \frac{kTt_{ox}}{q\Delta V_{GS}}\left[\ln(\tau_e/\tau_c)_{V_{BS}} - \ln(\tau_e/\tau_c)_0\right] \tag{2.14}$$

A procedure to extract the lateral trap position y_T was proposed by Amarasinghe *et al* [4, 20], based on the asymmetry of the time constant ratio as a function of the drain bias, represented in figure 2.3. Using the maximum ratio in forward (F) and reverse (R) biasing yields

$$\ln\left[\frac{\left(\frac{\tau_c}{\tau_e}\right)F_{max}}{\left(\frac{\tau_c}{\tau_e}\right)R_{max}}\right] = \frac{q}{kT}\frac{x_T}{t_{ox}}\left[\frac{y_T}{L}V_{DSFmax} - \left(1 - \frac{y_T}{L}\right)V_{DSRmax}\right] \tag{2.15}$$

with V_{DSFmax} and V_{DSRmax} the drain voltages corresponding to the maximum mark space ratio in forward and reverse operation, respectively. Equation (2.15) again assumes a negligible Coulomb blockade effect (negligible image charge on the gate and in the depletion region) and no impact of self-heating [8, 9]. The latter is a reasonable assumption for RTSs occurring around room temperature. Once the trap position has been determined from the mark space ratios, one can derive the trap level in the oxide from equation (2.12), whereby in general $E_F = E_{Fn} = E_{Fp} - qV_c$ and $\Psi_s = \Psi_{s0} + V_c$ [4]. The channel potential at the lateral trap position y_T can be approximated by $y_T V_{DS}/L$ for not too short channels and not too high lateral fields. Based on this analysis, it has been concluded that the energy of the RTSs at room temperature (normally repulsive or acceptors in nMOSFETs) are slightly above E_C, i.e. $E_{Cox} - E_T < \phi_e = 3.13$ eV [4].

While most of the RTSs appear to be located at some distance from the interface (typically 1–2 nm), the observation of fast RTSs at the interface has also been reported [21]. As expected, they correspond to faster switching events and smaller time constants (figure 2.10). The amplitude appears to be of similar magnitude as for slow traps [20], but multi-level transitions appear to be more frequent, due to the higher density of interface traps (D_{it}) close to the band edges. The time constants of figure 2.10 can be described by the SRH theory, with τ_e nearly independent of V_{GS} and an activation energy independent of E_F (or V_{GS}), i.e. $d(E_T - E_F)/dV_{GS} = 0$ and, hence, $x_T = 0$ (equation 2.13). The capture time is also several orders higher for slow oxide traps than for interface states and corresponds to a much smaller capture cross section. In this context, one often uses the approximation that $\sigma_n = \sigma_{n0}\exp(-x_T/\lambda)$, with σ_{n0} the electron capture cross section at the interface and λ the tunneling parameter (0.1 nm for the Si/SiO_2 interface) (see [22], for example). While this explains qualitatively the higher τ_c for deeper traps in the oxide, it has been found in practice that there is no correlation between the trap depth from the interface and the capture time constant for a large number of RTSs observed in 40 nm CMOS transistors [23]. As shown above, the capture cross section depends heavily on the

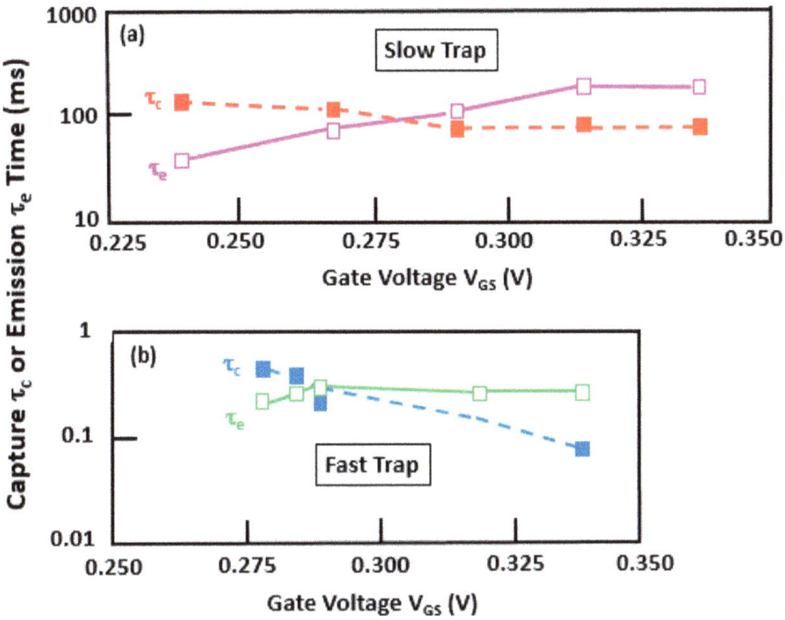

Figure 2.10. The average capture time (*a*) and the average emission time (*b*) as a function of gate voltage for a slow oxide trap and a fast interface state. (After Tsai *et al* [21].)

activation energy, which can take on a wide range of values at each trap depth. The limited dynamic range of time domain measurements selects only a restricted range of combinations of (σ_0, E_B) pairs.

Note that single-interface traps have also been studied using the charge-pumping technique at a frequency of 1 MHz [24]. Moreover, it has been reported that in conditions where electron valence band tunneling occurs, RTSs can be generated by electron–hole recombination [25–27]. RTS has also been observed in SOI MOSFETs [28, 29] and in some cases even in the back-channel, caused by traps in the buried oxide (BOX) [30–31]. The distinction between front- and back-channel traps can be achieved using a front-gate accumulation bias and a back-gate inversion bias.

In addition to the channel current, RTS can be found in the GIDL regime, with the gate biased in accumulation so that the *majority carrier (hole) transfer* is at its origin [32]. In that case, the drain current is governed by electron–hole pairs generated in the high-electric-field region in the drain overlapped by the gate. The generated minority carriers (i.e. holes in the n^+ drain) move to the body along the surface of the drain. A hole can be captured inside a trap in the gate dielectric which exists in the overlap region [32]. So, a trap with a captured hole decreases the vertical gate oxide field in a localized surface area in an nMOSFET and thus decreases the GIDL current as a result. Therefore, the occupied state (emission time) corresponds to the low current level in the GIDL waveform and vice versa. This is illustrated in more detail in figure 2.11 [32]. It is clear that, in this case, oxide traps at the complementary band edge of silicon are responsible for the RTS, so that the GIDL

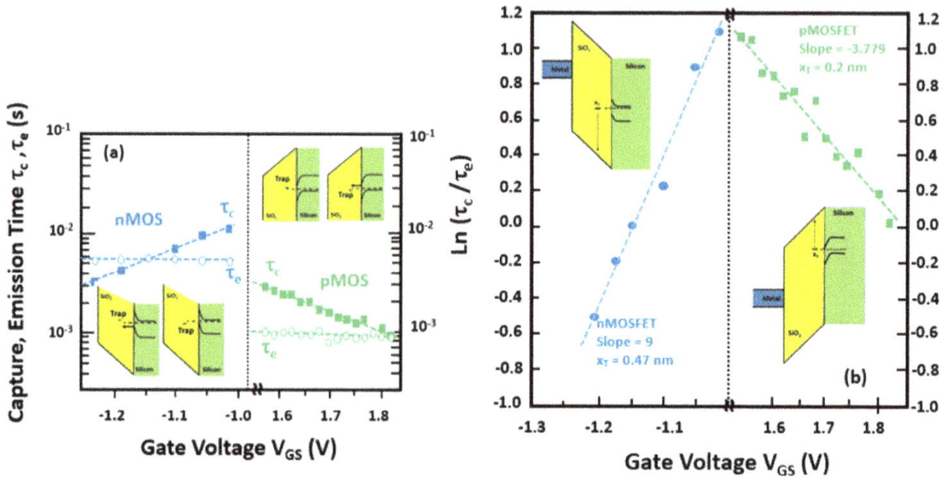

Figure 2.11. The lThe left and right sides of panel (*a*) show the dependence of τ_c and τ_e on the gate voltage in n- and pMOSFETs, respectively. The insets on the left and right represent schematically the energy band diagram to explain the capture probability of a hole and an electron with the gate voltage, respectively. The left and right sides of panel (*b*) show the dependence of $\ln(\tau_c/\tau_e)$ on V_{GS} for the GIDL current in n- and pMOSFETs, respectively. The dashed lines in (*b*) represent linear fitting curves to obtain the depth (x_T) of the trap from the interface between the oxide and the drain. (After Lee *et al* [32].)

mode allows the study of a different kind of defect (in energy position) compared to the normal inversion mode of operation in the same device. Since the minority carriers (electrons in nMOSFETs) are swept to the substrate giving rise to a substrate current I_B, one generally finds that the RTS in the GIDL current is reflected in I_B as well, with the amplitude $\Delta I_B \approx \Delta I_D$ ($= \Delta I_{GIDL}$) [32].

A recent study investigated the RTS in the drain–substrate junction leakage at $V_{GS} = 0$ V and at 100–125 °C in 0.1 μm scale MOSFETs [33]. This is important to understand the variability in the junction leakage in the off-state of a transistor. It has been shown that two types of bulk defects exist, based on the analysis of the activation energy of the RTS amplitude [33]: one which is not bias-dependent and another which causes an increase of the amplitude distribution with higher substrate doping density. Some of the defects studied also showed bistable behavior [33].

2.1.3 Non-SRH behavior: Coulomb blockade effects

Based on the SRH model of equation (2.2) it is expected that the average capture time of an RTS decreases with $\sim 1/I_D$. In practice, however, this is not always found, as exemplified by figure 2.12 [34], showing a superlinear decrease of τ_c with I_D. In other cases, even a reversal of the capture time with I_D (an increase) is found, as in figure 2.13 [35]. This is clearly outside the validity range of the standard SRH theory for RTS.

These observations have been interpreted in terms of the so-called Coulomb blockade energy, related to the image charge Q_G on the gate electrode, in the inversion layer and in the silicon depletion region [5, 19, 35]. The Coulomb energy

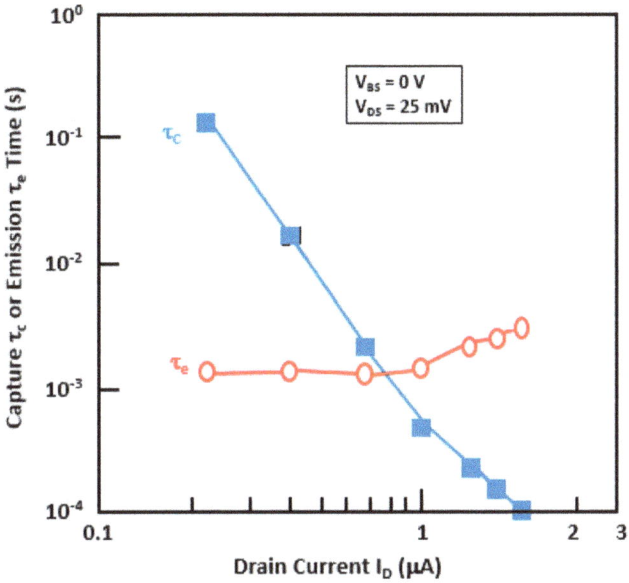

Figure 2.12. The dependence of the average capture (τ_c) and emission (τ_e) time constant of the RTS noise on the drain current at $V_{BS} = 0$ V and $V_{DS} = 25$ mV. (Reproduced with permission from Lukyanchikova *et al* [34]. Copyright 1998 AIP Publishing.)

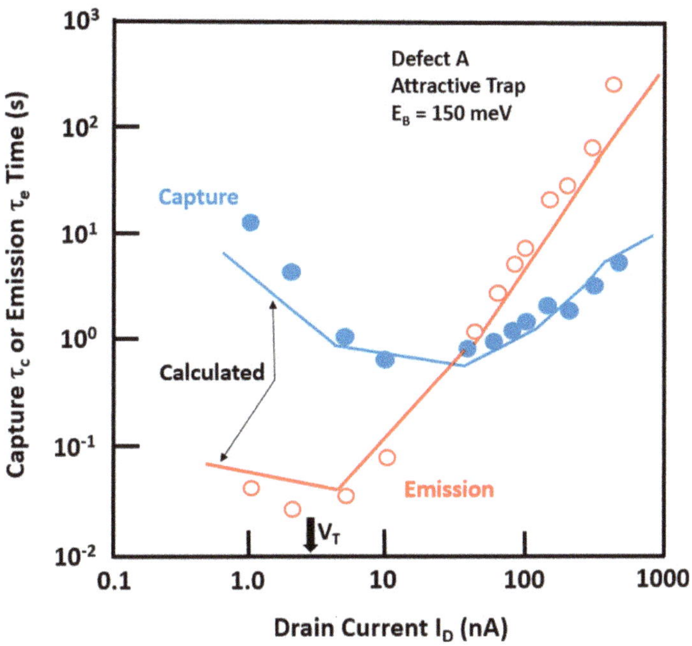

Figure 2.13. Capture and emission time constants of a single individual interface trap, measured as a function of the drain current I_D at $V_{DS} = 10$ mV by RTS. The continuous lines are calculated for an attractive trap with Coulomb energy $\Delta E(n_s)$. Only two parameters ($E_B = 150$ meV and $v_{thn}\sigma_n = 2 \times 10^{-18}$ cm^3 s^{-1}) are used to fit the magnitude of the time constants. (After Schulz [35].)

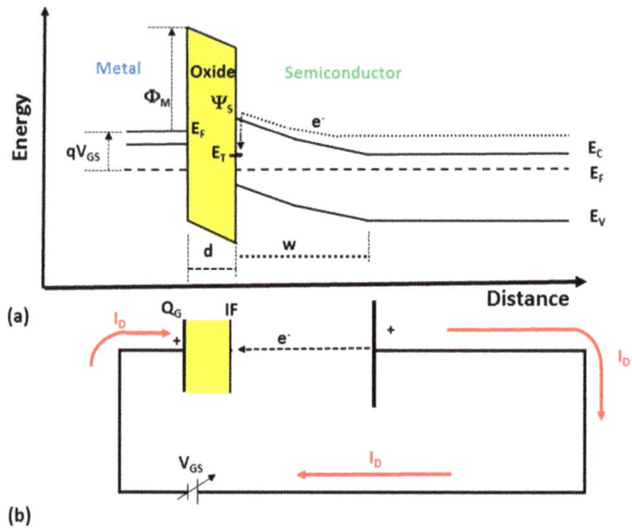

Figure 2.14. (*a*) Energy band diagram and (*b*) capacitance of a MOS structure with a bias voltage V_{GS} applied to the gate. An image charge Q_G is created on the gate electrode during the transfer by a displacement current I_d. The position IF indicates the location of the semiconductor/insulator interface (after [35]).

comes from the lateral transfer of the electron charge from the semiconductor to the trap location, which charges the capacitor by a displacement current I_d (figure 2.14). A free energy change of the capacitor electrode system must be taken into account in the capture and emission process into or out of traps located in space-charge regions. This free energy change strongly changes the trapping rate. In this context, a trap center may be viewed as a small capacitor structure of nanometer size, so that large local Coulomb forces (of order 100 meV) are possible upon charging with an electron [35]. This is similar to the charging of the capacitance of a single-electron transistor (SET).

The transfer of the charge of the trap is electrically visible in the measurable image charge Q_G which is created by the displacement current in the outer circuit of figure 2.14(*b*). For a trap right at the interface, we have [35]

$$\frac{Q_G}{q} = \frac{w/\varepsilon_S}{w/\varepsilon_S + t_{ox}/\varepsilon_{ox}} \tag{2.16}$$

with w the width of the depletion region in the silicon substrate, and ε_S and ε_{ox} the dielectric constants of silicon and SiO_2, respectively. The resulting Coulomb blockade energy $Q_G V_{GS}$ for a MOS capacitor reaches a maximum at the onset of inversion in figure 2.15 [35]. Beyond that point, the image charge on the gate becomes screened by the mobile charge density in the inversion layer, lowering the Coulomb energy E_{Coul}. Note that the energy ΔE_{Coul} in figure 2.15 can easily reach a few 100 meV.

The capture and emission rates are modified by the Coulomb energy according to the scheme of figure 2.16(left) for a neutral (acceptor) trap, switching between

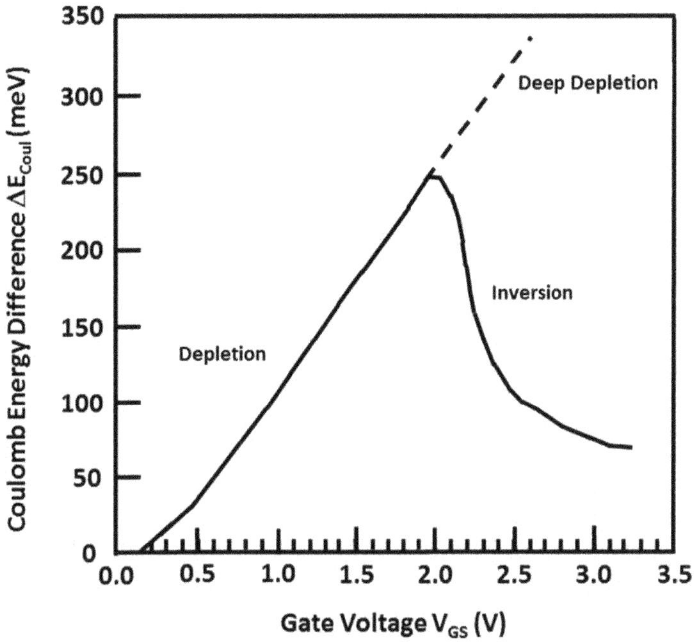

Figure 2.15. Calculation of the Coulomb energy difference ΔE_{Coul} at the interface of a MOS structure at room temperature as a function of applied voltage V_{GS}. The parameters are oxide thickness $t_{\mathrm{ox}} = 17$ nm, semiconductor doping $N_{\mathrm{a}} = 3 \times 10^{17}$ cm^{-3} and flat-band voltage $V_{\mathrm{FB}} = 0$ V. The offset by the band gap energy E_{G} for screening with holes is corrected so that ΔE approaches zero with screening. (Reproduced with permission from Schulz [35]. Copyright 1993 AIP Publishing.)

0 and – in an n-channel device and the scheme of figure 2.16(right) for an attractive (donor) trap switching between + and 0 by trapping of an electron. This can be generalized in a straightforward manner to the case of a pMOSFET.

In the case of *a repulsive trap*, the time constants become [40]

$$\frac{1}{\tau_{\mathrm{c}}} = v_{\mathrm{thn}}\sigma_{\mathrm{n}}N_{\mathrm{C}}\exp\left(-\frac{(E_{\mathrm{C}} - E_{\mathrm{T}})}{kT} - \frac{\Delta E_{\mathrm{Coul}}}{kT}\right) \tag{2.17a}$$

$$\frac{1}{\tau_{\mathrm{e}}} = v_{\mathrm{thn}}\sigma_{\mathrm{n}}N_{\mathrm{C}}\exp\left(-\frac{E_{\mathrm{b}}}{kT}\right) \tag{2.17b}$$

$$\tau_{\mathrm{e}}/\tau_{\mathrm{c}} = \exp\left(-\frac{(E_{\mathrm{T0}} + \Delta E_{\mathrm{Coul}} - E_{\mathrm{F}})}{kT}\right) \tag{2.17c}$$

In this case, the mark space ratio depends on the Coulomb energy, which is a function of the carrier density in the inversion layer. This is also reflected in the capture time, which exhibits an additional carrier density n_{s}-dependence, explaining the superlinear variation of τ_{c} on I_{D} [35].

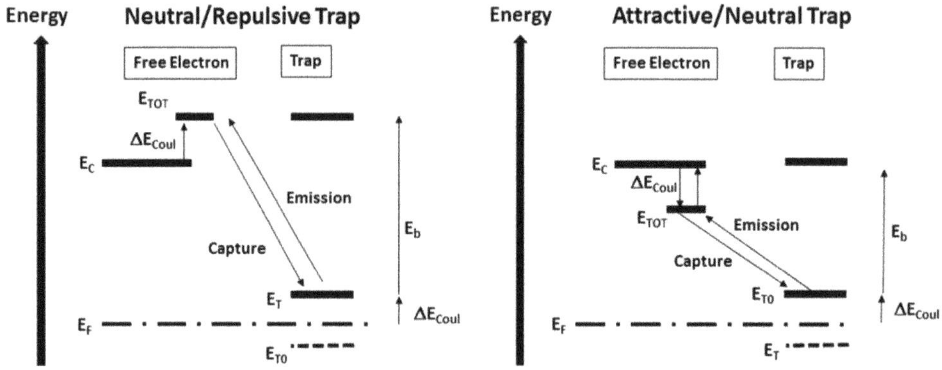

Figure 2.16. (Left) Energy-level scheme of an elementary charge in a space-charge region. The energy of an electron in the CB near the trap is shown on the left-hand side, the energy levels of a neutral (dotted bar E_{T0}) or repulsively (solid bar E_T) charged trap is shown on the right-hand side. E_b indicates the binding energy of the trapped electron. (Right) energy-level scheme of an elementary charge in a space-charge region as in the left panel. However, the charge states are attractive (dotted bar E_T) or neutral (solid bar E_{T0}) when the trap level is unoccupied or occupied, respectively, by an electron. (After Schulz [35].)

Following the scheme of figure 2.16 (right), the time constants for *an attractive trap* become [35]:

$$\frac{1}{\tau_c} = v_{thn}\sigma_n N_C \exp\left(-\frac{(E_C - E_T)}{kT}\right)\exp\left(+\frac{\Delta E_{Coul}}{kT}\right) \tag{2.18a}$$

$$\frac{1}{\tau_e} = v_{thn}\sigma_n N_C \exp\left(-\frac{\left(E_b - \Delta E_{Coul}\right)}{kT}\right) \tag{2.18b}$$

$$\tau_e/\tau_c = \exp\left(-\frac{E_{T0} - E_F}{kT}\right) \tag{2.18c}$$

In this case, the equilibrium occupation factor is not affected by the Coulomb energy (2.18*c*), as also verified experimentally in figure 2.17. This is related to the fact that the occupied trap is in a neutral charge state, thus not giving rise to an image charge on the gate. The amazing feature for an attractive trap is the turn-around behavior of τ_c (figure 2.13) that cannot be accounted for in the classical SRH framework. However, utilizing the variation of the Coulomb energy ΔE_{Coul} with gate voltage (or charge density in the channel), depicted in figure 2.14, readily explains this trend. The reduction of ΔE_{Coul} with increased screening decreases the shift of the trap level upon charging and diminishes thus the energy gain. As can be seen in figure 2.15, the emission time constant also becomes strongly gate-voltage-dependent in this case. It has furthermore been argued that for high Coulomb energies the thermal activation may not be sufficient to free the mobile carrier from the intermediate trapped state E_{TOT} in figure 2.16 into the CB. This may in particular be the case at low temperatures, where the carrier is bound to the defect in the band and also in the

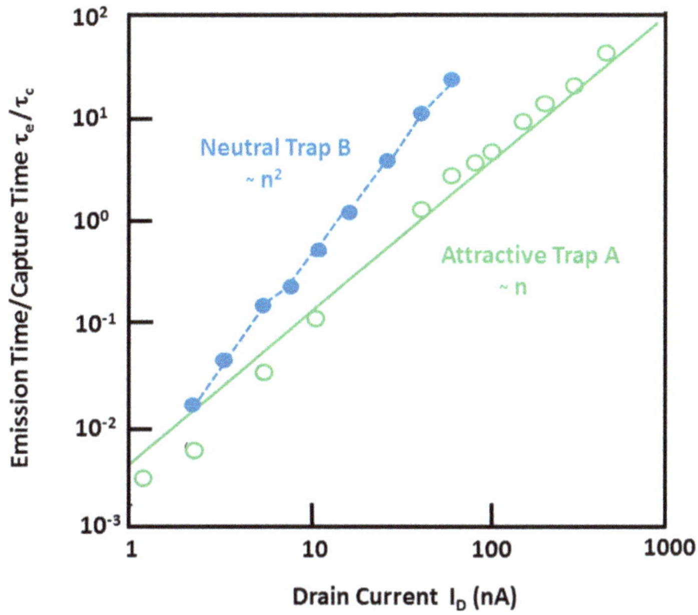

Figure 2.17. Ratio of the emission and capture time constants τ_e/τ_c as a function of the source–drain current I_D at $V_{DS} = 10$ mV. The dashed line for the neutral defect B connects the data points to guide the eye. The slope is approximately proportional to the square of the carrier density n^2. The solid line fit to the data points for the attractive defect A indicates a slope proportional to n as predicted by the SRH ratio. (Atfer Schulz [35].)

trapped state at E_{T0}. In other words, the carrier is bound in a bistable configuration, leading to anomalous switching [7, 12], whereby no carrier is exchanged with the channel and the RTS amplitude is only associated with scattering ($\Delta\mu$).

A procedure to extract ΔE_{Coul} has been developed by Mueller *et al* [5] and this issue will be discussed in more detail in chapter 3.

2.1.4 Tunneling transitions

It is evident from the above that at room temperature the time constants of typical RTSs are thermally activated. This implies that the transitions do not correspond to elastic tunneling at a constant energy around E_F, which is assumed in many cases. The McWhorter theory for $1/f$ noise is a notorious example, relying on elastic tunneling to and from oxide traps near the insulator/semiconductor interface. The necessary range of time constants then directly follows from the fact that traps are located at different depths; the deeper traps corresponding to a larger tunneling time, according to $\tau = \tau_0 \exp(x_T/\lambda)$. Based on this theory, a $1/f$ noise spectrum can readily be transformed into an oxide trap (border trap) density profile. However, as pointed out recently by Campbell *et al* [36], this may be incorrect, especially for thin gate oxides, where the elastic tunneling times are smaller than necessary to explain the observed RTSs. A similar remark may be valid for the modeling of retention and charging times of MOS-based memory devices, as will be discussed in chapter 5.

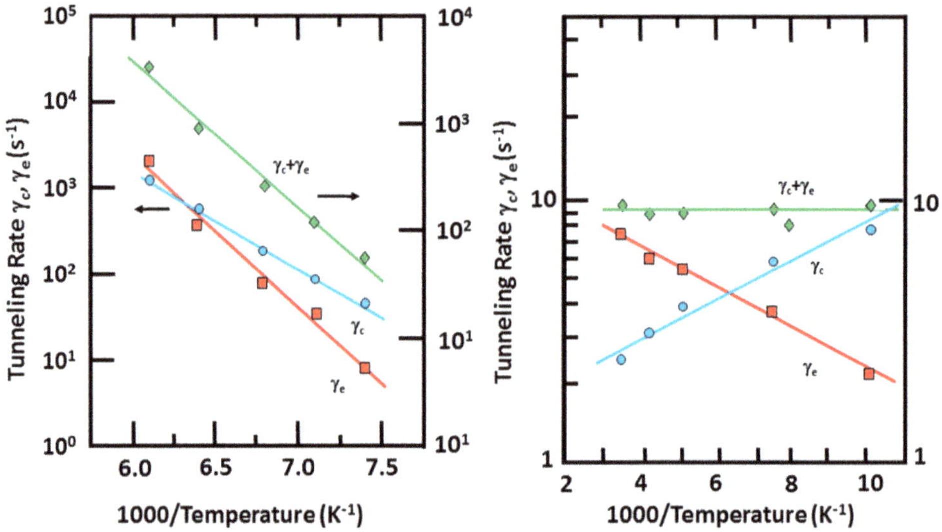

Figure 2.18. The temperature dependence of the capture, the emission and the total switching rate for two different RTSs in an SOI nMOSFET at 298 K and $V_{DS} = 5$ mV. (Reproduced with permission from Shi *et al* [38]. Copyright 2001 IOP Publishing.)

Some experimental support for the McWhorter elastic tunneling model for noise has been derived from the low frequency (LF) noise spectra of MOSFETs with planar layers of Si nanocrystals embedded in an SiO_2 gate at a certain distance from the interface [37]. These nanocrystals can be considered as an array of giant traps capable of exchanging charge with the channel. Tunneling lengths in the range of 0.086 to 0.093 nm have been obtained, in good agreement with the value of 0.1 nm usually assumed for λ.

Evidence that some RTSs rather show no thermally activated tunneling transitions has been given on a few occasions [38, 39]. The likelihood for finding such an RTS becomes higher at lower temperatures, where less thermal energy is available for activation of the charge transitions. Figure 2.18 compares the Arrhenius diagrams for a standard thermally activated RTS in an nMOSFET and a tunneling trap [38]. The major difference is that the total transition rate $\gamma_e + \gamma_c$ is fairly temperature-independent in the latter case. Note also the opposite temperature behavior of γ_c in figures 2.18(left) and 2.18(right), strongly suggesting the attractive nature of the tunneling trap (donor). Another feature is that the amplitude of the second trap is 70% compared with a few percentage for the ordinary thermally activated RTSs [38]. The main origin of this large ΔI_D is $\Delta\mu$ fluctuation (scattering).

The temperature independence of the total switching rate can be explained by considering the tunneling rates between a localized state and a two-dimensional electron gas (2DEG) [38], dictated by Fermi's golden rule

$$\gamma_c = \frac{2\pi}{\hbar} D\Delta^2\left[1 - f(\epsilon)\right] \tag{2.19a}$$

$$\gamma_e = \frac{2\pi}{\hbar} D\Delta^2 f(\epsilon) \tag{2.19b}$$

$$\gamma_c + \gamma_e = \frac{2\pi}{\hbar} D\Delta^2 \tag{2.19c}$$

In equations (2.19) D is the density of electron states in the 2DEG (including quantization) and Δ is the tunnel matrix element; $\hbar$ is the reduced Planck's constant $h/2\pi$. In addition, $f(\in)$ is the Fermi function, given by:

$$f(\epsilon) = \frac{1}{1 + \exp(\epsilon/kT)} \tag{2.20}$$

The switching rate in equation (2.19*c*) is independent of temperature. This can be further worked out, resulting in [40]

$$\gamma_e = \gamma_{e0} \exp\left[-2\frac{\left(2m_{ox}(E_{Cox} - E_T)\right)^{1/2}}{\hbar} x\right] \tag{2.21}$$

with m_{ox} the electron effective mass in SiO_2, with γ_{e0} specified in [40].

A gradual change from thermally activated transitions to temperature-independent ones has also been observed in a single RTS at low T [39]. This is illustrated in figure 2.19 for a *repulsive* trap in a p-channel enhancement mode MOSFET with

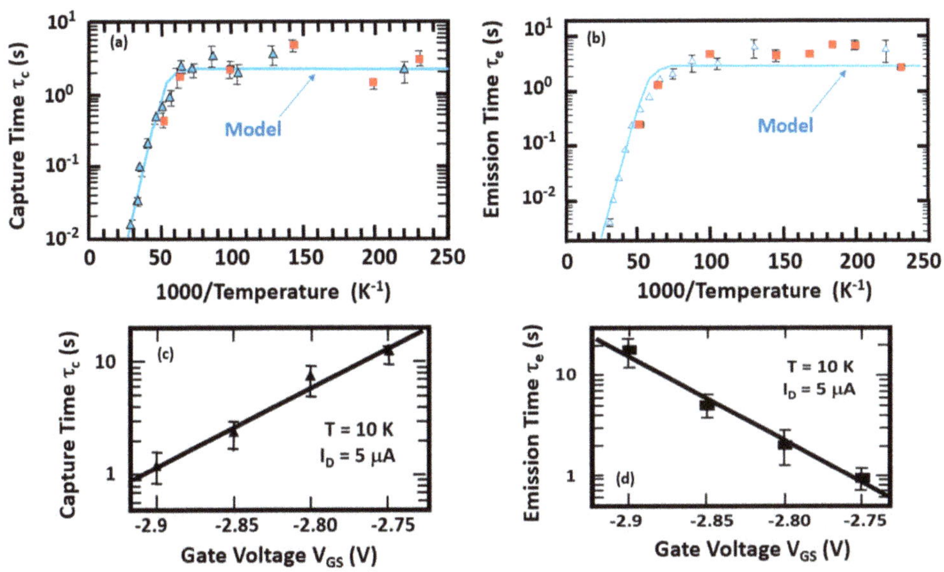

Figure 2.19. Arrhenius plot showing the temperature dependence of the mean capture (*a*) and emission time (*b*) for an RTS at 4.2 K–30 K. The open triangles represent data measured at $V_{GS} = -3.00$ V and $I_D = -15$ μA, while the solid circles represent data measured with $V_{GS} = -2.85$ V and $I_D = -5$ μA. The line represents the theory with three adjustable parameters obtained using a weighted non-linear least-squares fit to the triangle data. Insets show the strong voltage dependence of the RTSs (after Scofield *et al* [39]).

$t_{ox} = 18$ nm. Modeling the emission and capture time constants (or rates) by the sum of a temperature-activated and a temperature-independent part has been proposed, i.e. $1/\tau = 1/\tau_1 + 1/\tau_2$, with $\tau_1 = \tau_{10}\exp(E_a/kT)$ [39]. This yields a good fit with the data points in figure 2.19, resulting in small activation energies for τ_c (18 meV) and τ_e (19 meV).

Given the predominance of thermally activated border traps near the Si/SiO_2 interface, alternative models for the trapping time constants have been proposed. One interesting approach, stemming from the flash memory field, considers charge exchange by inelastic tunneling transitions [41, 42]. Briefly speaking, the capture rate for an electron is given by [42]

$$\frac{1}{\tau_c} = \frac{4\pi m_n kT}{h^3}\sigma_c \int D(E)\ln\left(1+e^{-\frac{E-E_F}{kT}}\right)dE \tag{2.22}$$

with m_n the sum of the effective masses for the six CB minima in the silicon substrate evaluated in the plane transversal to the silicon oxide direction and σ_c an effective capture cross section. $D(E)$ is the tunneling transparency and E is the carrier energy component in the direction normal to the interface [42]. Considering trap energies E_T well above the Fermi level E_F, equation (2.22) can be simplified to [42]

$$\frac{1}{\tau_c} \approx \frac{4\pi m_n (kT)^2}{h^3}\sigma_c D(E_T)e^{-\frac{E_T-E_F}{kT}} \tag{2.23}$$

Formally, this is equivalent to the thermally activated SRH equation (2.3), i.e. $1/\tau_c \sim 1/\tau_e\exp[-(E_T - E_F)/kT]$, with the emission rate given by the tunneling rate of equation (2.22). Under the Wentzel–Kramers–Brillouin (WKB) approximation, in the direct-tunneling regime the transparency can be approximated as [42]

$$D(E_T) = \exp\left[-\frac{4\sqrt{2m_{ox}}}{3\hbar qF}\left\{\left(E_T' + qFx_T\right)^{3/2} - \left(E_T'\right)^{3/2}\right\}\right] \tag{2.24}$$

with F the electric field in the oxide and E_T' the distance of the trap level to the SiO_2 CB edge ($E_{Cox} - E_T$). As the random telegraph noise (RTN) traps are located near the substrate interface $qFx_T << E_{Cox} - E_T$, equation (2.24) can be simplified to [4]

$$D(E)_T = \exp\left[-2\frac{\left(2m_{ox}(E_{Cox} - E_T)\right)^{1/2}}{\hbar}x_T\right] \tag{2.25}$$

Combined with equation (2.22), this is equivalent to replacing the capture cross section with $\sigma_0\exp(-x_T/\lambda)$ as mentioned above. Based on this approximation, a spectroscopic investigation of RTSs in a 65 nm Negative-OR technology was undertaken in [41] and [42].

At liquid helium temperatures, where the 2DEG is degenerate, the RTS study of tunneling events at the Si/SiO_2 interface enables some fundamental properties to be investigated [6, 43, 44]. In this way, a tunneling two-state system (TSS) has been identified at 4.2 K in HC stressed nMOSFETs [43], which is characterized by the fact that the (attractive) trap system only exchanges energy with the 2DEG and not carriers. The change in channel resistance is then created by a change in the scattering cross section for inversion-layer electrons. The behavior of the TSS can be completely characterized by the (small) energy separation between the two states, the renormalized tunnel matrix element Δ and a dimensionless coupling factor α_{coup} which depends on the nature and the strength of the interactions with the surrounding electron gas [43]. In this case, the switching rate $(\tau_c\tau_e)^{1/2}$ exhibits a resonance at a specific gate voltage V_{GSO} and the mark space ratio can be described by $\exp(-\Delta E/kT)$ with $\Delta E = -q\eta(V_{GS} - V_{GSO})$ and $\eta = x_T/t_{ox}$. This is equivalent to the energy of a dipole with magnitude $q\eta t_{ox}$ aligned parallel to the oxide electric field F. The stress-induced dissipative TSSs behave non-Poissonian-like, as shown in figure 2.20, and are characterized by a weak coupling constant α_{coup} in the range of 10^{-3}. The energy separation can be controlled by the gate voltage, independent of the applied magnetic field [43].

In the millikelvin range, RTSs provide a powerful tool for investigating tunneling between a 2DEG and a single-defect state [44]. It has been observed that the tunneling

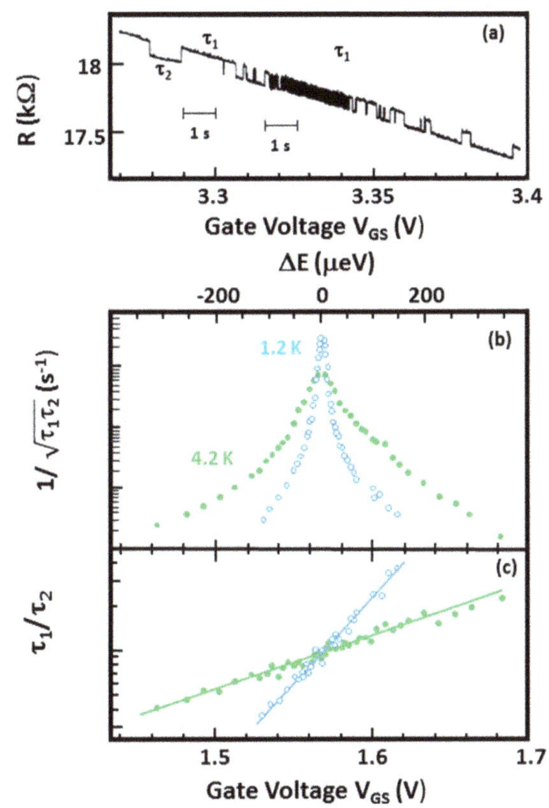

Figure 2.20. (*a*) Gate-voltage sweep at the indicated rate past a resonant RTS centered at V_{GSO} = 3.33 V at 4.2 K with a time resolution of 1 ms. (*b*),(*c*) Profile of RTS in weak inversion at two temperatures: o, 1.2 K; •, 4.2 K. The points at 4.2 K are offset in V_{GS} by +12 mV, to compensate for the small change in V_{GSO}. (Reproduced with permission from Cobden *et al* [43]. Copyright 1993 the American Physical Society.)

rate shows a peak when a defect level lines up with the Fermi energy, in excellent agreement with the theory of the Fermi-edge singularity at finite temperature.

2.2 RTS amplitude behavior

The amplitude of a two-level RTS is a well-defined quantity, without statistical distribution, that can be easily evaluated for not too small signals. It turns out to be far more difficult to understand its dependence on gate and drain bias and, in particular, the large spread which has been found from sample to sample and even for different RTSs in the same transistor, as illustrated by figure 2.21 [18]. As indicated in figure 2.21(*a*), the normalized amplitude is significantly larger than what can be expected from a simple number fluctuation model, corresponding to a normalized amplitude $\Delta I_D/I_D$ of $1/N$ in strong inversion. The first RTS amplitude model was worked out by Uren *et al* [45], assuming a uniform channel, whereby the capture of an electron by a trap locally enhances the resistivity R, resulting in

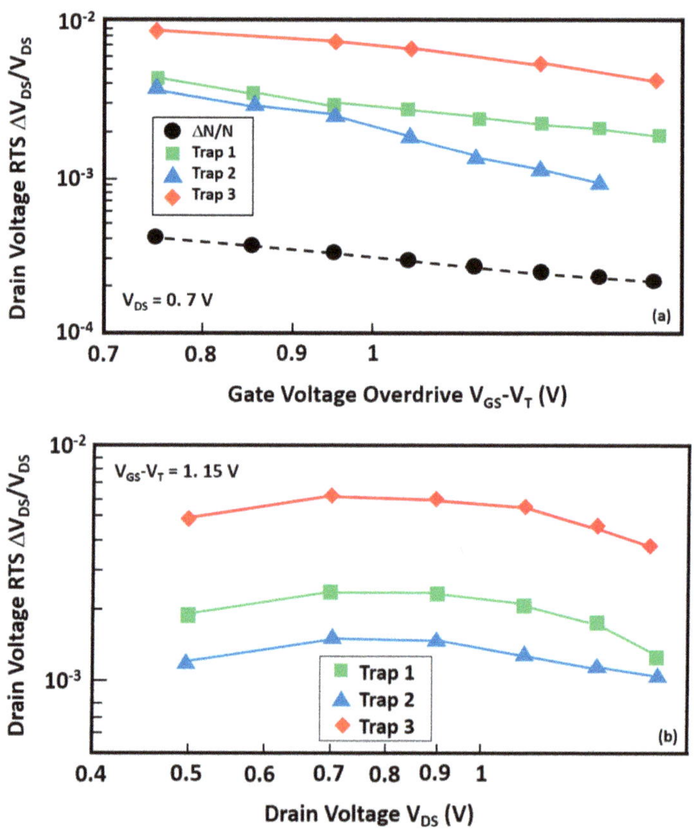

Figure 2.21. The dependence of the normalized drain voltage RTS on (*a*) the effective gate voltage at V_{DS} = 0.7 V (also compared to the calculated number fluctuations, dashed line) and (*b*) the drain voltage at ($V_{GS} - V_T$) = 1.15 V for three signals in a nMOSFET in strong inversion. (Reproduced with permission from Amarasinghe and Çelik-Butler [18]. Copyright 2000 Elsevier.)

$$\frac{\Delta R}{R} = \frac{1}{WL} \frac{q\beta}{\left(C_{ox} + C_D - \beta Q_n\right)} \tag{2.26}$$

with $\beta = q/kT$, C_{ox} the oxide capacitance per unit of area, C_D the depletion capacitance per unit of area and Q_n the channel charge density in C cm^{-2}.

From this it follows that the normalized RTS amplitude will increase with the inverse device area and in strong inversion approaches q/WLQ_n or $1/N$, with N the total number of carriers in the channel. While equation (2.26) qualitatively explains the amplitude dependence on V_{GS}, resulting in a $1/I_D$ roll-off in strong inversion [45, 46], it also predicts a single value which is not found in practice (see e.g. figure 2.21). Even in a worst case scenario, the number fluctuations can only explain an amplitude not larger than 1% in weak inversion [47, 48] and can thus not explain the large RTS amplitudes frequently observed [49]. In addition, the proximity of the gate in thin-oxide devices will result in an amplitude which becomes smaller than one elementary charge in weak inversion, due to the image charge screening effect.

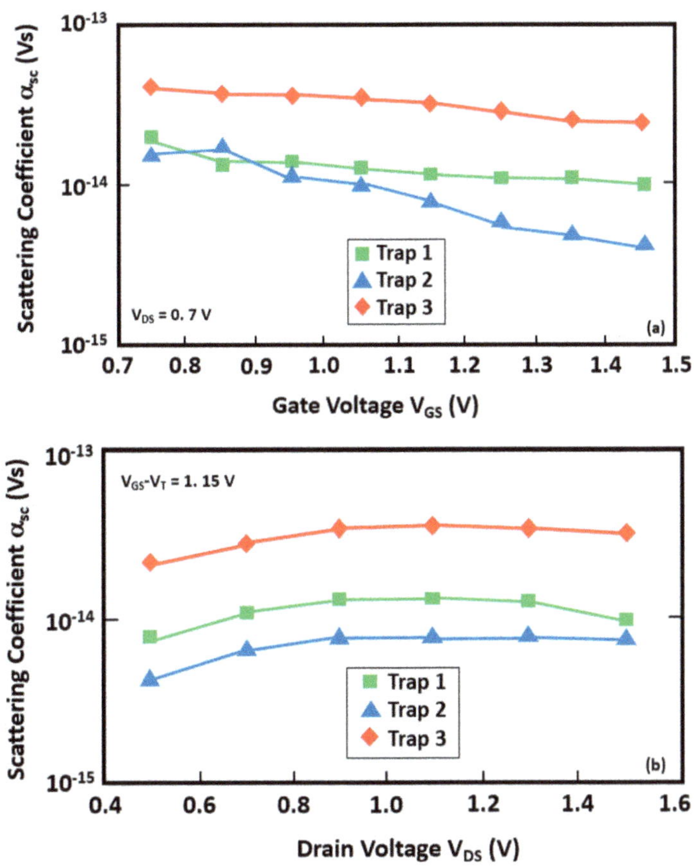

Figure 2.22. The scattering coefficient dependence on (*a*) drain voltage at $V_{GS} - V_T = 1.15$ V and (*b*) effective gate voltage at $V_{DS} = 0.7$ V for three oxide traps. (Reproduced with permission from Amarasinghe and Çelik-Butler [18]. Copyright 2000 Elsevier.)

Of course, when an oxide trap changes its charge state upon carrier capture, its scattering cross section also changes. This means that trapping is expected to be associated with a change in the mobility $\Delta\mu$, given by $\pm\alpha_{sc}\mu/WL$ [8, 50]. The total amplitude, assuming a uniform channel and a uniform change in current then becomes [8, 50]

$$\frac{\Delta I_D}{I_D} = \frac{\Delta n_s}{n_s} + \frac{\Delta\mu}{\mu} = -\frac{1}{WL}\left[\frac{1}{\frac{kT}{q(C_{ox}+C_D)} + n_s} \pm \alpha_{sc}\mu\right] \tag{2.27}$$

In equation (2.27) $Q_s = qn_s$, with n_s the surface carrier density in cm^{-2}.

It has frequently been demonstrated that the $\Delta\mu$ term dominates the Δn contribution [18, 51]. Detailed calculations demonstrate that the normalized amplitude remains constant as long as the depletion charge is higher than the inversion charge density at room temperature [51]. At 88 K, the amplitude roll-off sets in earlier. From the $\Delta\mu$ contribution, one can derive the scattering coefficient α_{sc} defined in equation (2.27). According to figure 2.22, this parameter is in the range of 10^{-14} Vs for the Si/SiO_2 system and decreases with increasing gate voltage according to the empirical relationship [18]

$$\alpha_{sc} = K_1 + K_2 \ln(n_s) \tag{2.28}$$

with K_1 and K_2 constants associated with the fitting of the experimental data. However, a simple estimate based on a scattering length of 10 nm indicates that the $\Delta\mu$ term is at most 1% [48]. This can by no means explain the high relative amplitudes up to 10% reported in figure 2.23 [52, 53]. In fact, even giant RTSs of several tens of per cent have been reported [54–56], which clearly fall out of the range that can be explained by equation (2.27). In order to model such amplitudes (and the distribution reported in figure 2.23), one has to develop more sophisticated models, assuming first of all a non-uniform, local impact of the charged trap on the current and, in addition, considering a non-uniform channel in a random surface-potential landscape. This will be discussed in the following chapter on modeling.

Most studies on the RTS amplitude have focused on its gate voltage (drain current dependence), but the drain voltage dependence may also reveal useful information [18, 46, 57–59]. In particular the asymmetry of the amplitude when exchanging source and drain contacts can be helpful in unraveling more detailed trap parameters [46, 57, 58]. A method has been proposed, based on the asymmetry of the amplitude and a model for the channel current, to derive the lateral trap position [46, 57] and eventually using RTS as a local atomic probe of the surface potential along the channel. The problem with such an approach is that in state-of-the-art devices, the channel doping is far from uniform and randomly distributed. This implies that such a method cannot be treated by simple analytical formulas and requires more sophisticated 2D and 3D so-called 'atomistic' numerical simulations (see chapter 3). In addition, the occurrence of RDFs and short-channel effects may affect the RTS amplitude so strongly that it is no longer possible to extract meaningful data on the trap location (see the following section).

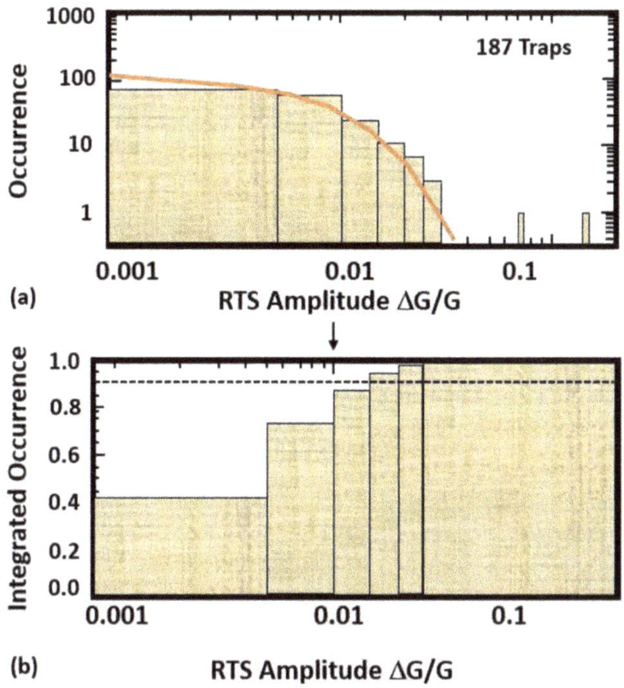

Figure 2.23. Histogram of 187 experimentally observed RTS amplitudes $\Delta G/G$ in sub-micrometer MOSFETs. (*a*) The histogram columns depict the occurrence number. The solid curve represents the distribution calculated under the assumption of a Gaussian current density distribution in the MOSFET channel with expectation value 0.064 and standard deviation 0.042. (*b*) The histogram columns depict the integrated occurrence. The dotted line marks the 90% level, the arrow indicates the modulation value $\Delta G/G = 1\%$. (Reproduced with permission from Mueller and Schulz [52]. Copyright 1997 AIP Publishing).

A final issue is the impact of device scaling on the drain current RTS amplitude. According to equation (2.27), the relative amplitude is proportional to the inverse of the transistor area $(WL)^{-1}$ and should thus increase for smaller devices. The length dependence of the RTS amplitude ΔI_D has been verified experimentally using a dedicated test structure [61, 62]. In ultra-narrow channel transistors, huge RTSs have been observed [54, 62, 63], which points to the importance of the trap location with respect to the current path: a trap in a strategic position can nearly completely block the channel current in a minimum surface-potential path or in a geometrically constricted, narrow current filament. In addition, when scaling the gate oxide thickness, screening by the (metal) gate becomes more and more important. The image charge induced on the gate electrode modifies the RTS amplitude drastically, especially in weak inversion. A simple analytical model has been worked out by Zhang *et al* to estimate the impact of the gate proximity [64]. The cored out area of a trap can be derived from

$$\frac{q}{4\pi\epsilon\sqrt{\frac{L_t^2}{\pi} + (x_T + t_{qm})^2}} - \frac{q}{4\pi\epsilon\sqrt{\frac{L_t^2}{\pi} + \left[2(t_{ox} + t_{dp}) - x_T\right]^2}} = \frac{kT}{q} \qquad (2.29)$$

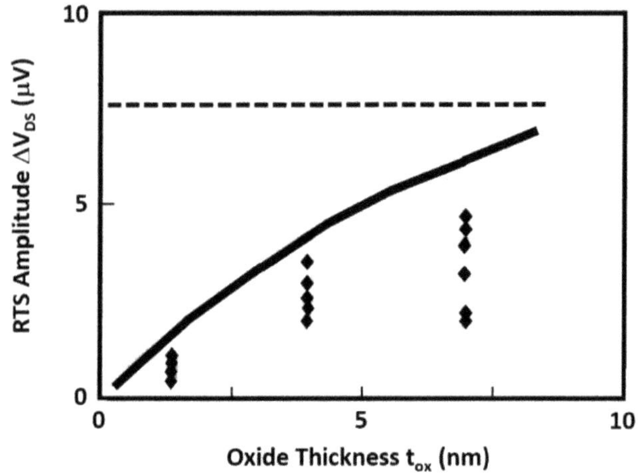

Figure 2.24. The dependence of the RTS amplitude on t_{ox}. (Reproduced with permission from Zhang *et al* [64]. Copyright 2008 IOP Publishing.)

with L_t the side of a square-shaped area around the trap where the current is blocked when occupied by a carrier (see chapter 3 for more details). In equation (2.29) t_{qm} is the offset of the carrier centroid from the interface due to inversion-layer quantization and t_{dp} is the polysilicon depletion thickness (equal to 0 for a metal gate). As can be derived from figure 2.24, the RTS amplitude will be significantly reduced by the image charge effect when scaling the oxide thickness. This also results in a reduction of the corresponding LF noise PSD.

2.3 RTS in the gate current of a MOS device

The first observations of RTS noise in thin oxides were obtained from scanning tunneling spectroscopy [65] and from small-area metal–oxide–semiconductor tunnel diodes [66]. An overview of these early results can be found in [67]. In this section, the focus is on process-induced RTSs. The occurrence of RTSs and more complex multi-level fluctuations in thin oxides after electrical stress will be discussed in more detail in chapter 5. More recently, interest in RTS in the gate current I_G has revived [68–73], since thin oxides are now implemented in deep submicron CMOS technologies. In addition, the gate-current (I_G) RTN has also been analyzed in high-κ transistors with a thin SiO_2 interfacial layer (IL) [74–77]. One of the main conclusions is that the gate current is more sensitive to oxide traps than the channel current, yielding larger relative amplitudes (figure 2.25). Avellan *et al* have pointed out that the gate-current change induced by the same trap has opposite sign compared to the drain current amplitude, while the switching events in the drain and gate current are correlated (figure 2.25) [68]. In other words, the gate current is higher when the trap is occupied by an electron. A possible explanation is that the image force reduction induced by the trapped electron on the tunneling barrier enhances the transmission probability across the gate dielectric.

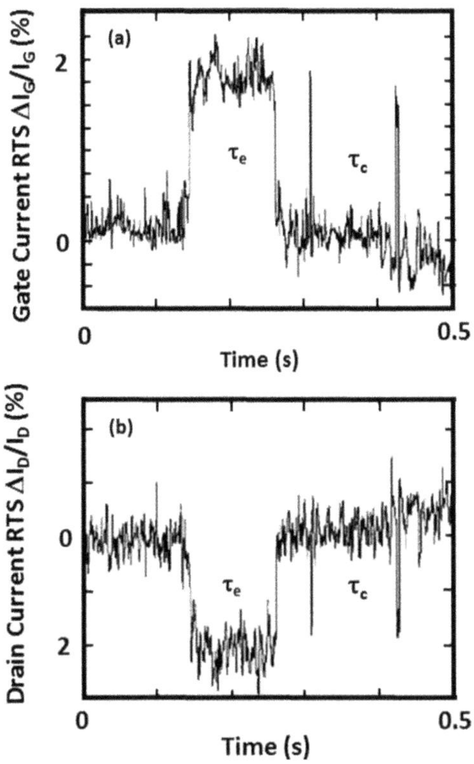

Figure 2.25. Typical RTS in the time domain in gate (*a*) and drain (*b*) currents at $V_{GS} = 0.45$ V and $V_{DS} = 50$ mV for a device with $L = 0.5$ μm and $W = 10$ μm, $t_{ox} = 1.3$ nm. (Reproduced with permission from Avellan *et al* [68]. Copyright 2001 AIP Publishing.)

Further studies also indicated that the emission and trapping time constants of gate-current RTS may behave differently than the usual channel RTS (see figure 2.26), in particular if they correspond to a (soft) breakdown (BD) spot [70]. While the upper figure illustrates the usual trends, with τ_c decreasing with gate voltage and crossing the more or less constant τ_e at some point, in the lower diagram one can see that for certain signals both time constants decrease with increasing V_{GS}. In order to explain the observations, multi-phonon capture was assumed [70], yielding a thermally activated capture time constant, as in equation (2.9) for a drain current RTS. At room temperature the capture probability is then given by [70]

$$P_c \propto \exp\left(-\frac{(-S_0\hbar\omega + E_C - E_T)^2}{4S_0\hbar\omega kT}\right) \tag{2.30}$$

Equation (2.30) assumes strong lattice coupling of the defect through the Huang–Rhys factor S_0. This parameter is a measure for the number of phonons of energy $\hbar\omega$

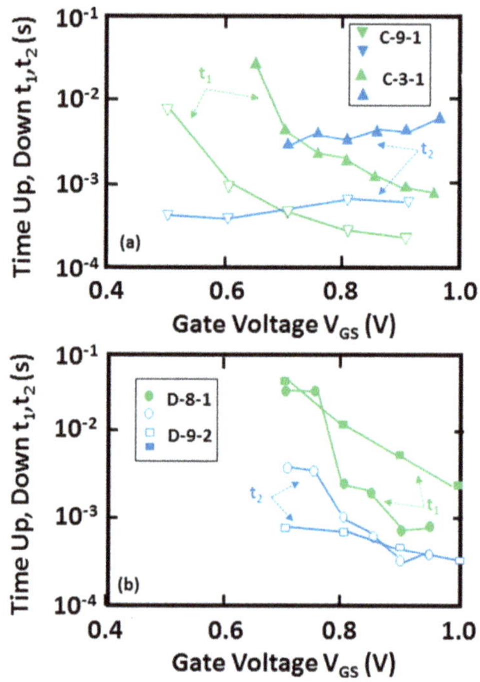

Figure 2.26. Measured up (closed symbols) and down times (open symbols) as a function of gate voltage. (*a*) Classical behavior exhibiting a crossover and (*b*) decreasing up and down times. (After Avellan *et al* [70].)

emitted upon capture of an electron by the trap. For strong lattice coupling, $E_C - E_T < S_0\hbar\omega$, so that we have

$$P_c \propto \exp\left(-\frac{S_0\hbar\omega}{4kT} + \frac{E_C - E_T}{2kT}\right) \tag{2.31}$$

which is equivalent to equation (2.9). The different behavior reported in figure 2.26 has been explained by considering that for a trap in a BD spot, the Fermi level is changed from the silicon one to the one corresponding with the gate. This modifies the emission time constant, so that a reduction with increasing V_{GS} is obtained [70].

Based on the sign of the slope of $\mathrm{dln}(\tau_c/\tau_e)/\mathrm{d}V_{GS}$, it has been shown that in the case of high-κ gate stacks one can make the distinction between an I_G trap in the high-κ oxide communicating with the channel or with the gate electrode [74–77]. This is schematically represented in figure 2.27, defining the parameters for the type I (channel) and type II (metal gate) traps. In the former case, the slope is negative, while it is positive in the second case. In other words, when the gate voltage increases, the probability of electron trapping decreases [74].

Recently, several methods have been proposed to employ gate-current RTS and $1/f$ noise instead of their drain current counterparts for trap analysis in advanced gate stack devices, with more or less success [76–83]. For example, Crupi *et al* have

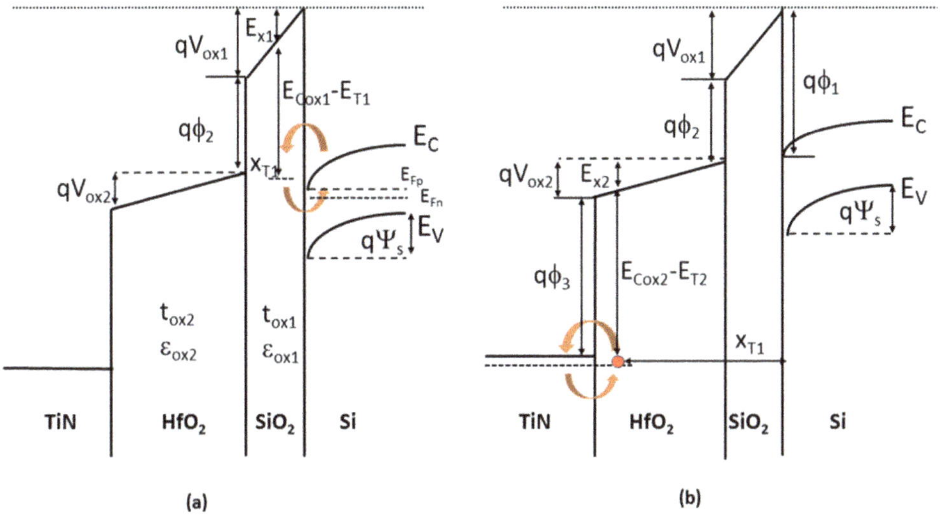

Figure 2.27. (*a*) Energy band diagram of the high-*κ* dielectric device. Energy equations for the trap interacting with the channel are derived from the difference between the trap energy level and bulk electron Fermi level. (*b*) Energy band diagram of the high-*κ* dielectric device when the trap interacts with the gate electrode. (After Lee *et al* [75].)

proposed a semi-empirical model, relating the oxide-trap density to the gate-current noise spectral density S_{IG} [79]:

$$N_T(E_F) = \frac{AfS_{IG}}{a_0^2 \lambda I_G^2 kT} \quad (\text{in cm}^{-3}\,\text{eV}^{-1}). \tag{2.32}$$

This is based on an SRH type approach assuming single-electron capture by tunneling, neglecting emission. The empirical factor a_0, corresponding with the one-electron blocking area, considers that when the trap is charged, no electron current can pass through it [83]. It is estimated theoretically as follows

$$a_0 = \int_0^{\infty} 2\pi r \left[1 - \frac{T(r)}{T_0}\right] \mathrm{d}r \tag{2.33}$$

The tunneling probability $T(r)$ has been calculated for the neutral trap (T_0) and when the center is charged. In the latter case, the effect of the trap potential $q\phi(y, r)$ on the barrier height, including image charge effects, has been accounted for using a semi-analytical approach [79].

As shown in figure 2.28, the blocking radius decreases with the dielectric constant κ, increases with the oxide thickness (less gate screening) and is almost independent of the gate voltage. The reduction of the blocking radius for thinner oxides clearly emphasizes the importance of the image charge and is equivalent to a significant reduction of the number contribution to the RTS amplitude below $1/N$, in agreement with figure 2.24 [64]. It also explains why the oxide-trap density derived from channel $1/f$ noise may become underestimated [72]. Nevertheless, using the gate-current $1/f$ noise, oxide-trap densities in good agreement with the values derived

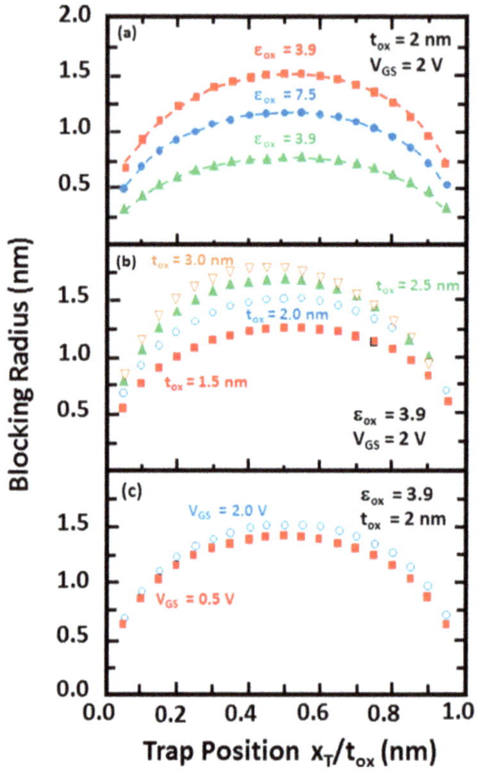

Figure 2.28. The radius of the one-electron blocking area decreases with the dielectric constant (*a*), increases with the oxide thickness (*b*), and is almost independent on the gate voltage (*c*). In all cases the radius shows its maximum for a trap in the middle of the oxide, i.e. $x_T = t_{ox}/2$. (Reproduced with permissions from Crupi *et al* [79]. Copyright 2009 AIP Publishing.)

from the drain current noise spectral density S_{ID} have been obtained [68]. However, when evaluating the cored out area from gate-current RTS, much larger values than theoretically predicted have been obtained (figure 2.29). This indicates that a_0 is at best an empirical parameter and questions also the validity of the one-electron picture in the case of gate-current RTSs [66, 67], in particular for stress-induced two-level fluctuations (TLFs).

In this context, it should be remarked that an alternative view exists on the origin of TLFs in a MOS diode [67], which is compared to the Kirton–Uren picture for channel RTSs in figure 2.30. In the latter case, also described above in figure 2.8 [15], the capture of carriers is thermally activated, because a barrier has to be overcome elastically (by absorbing vibrational energy from the lattice) in order to distort the defect configuration (arrangement) in such a way that filling by direct elastic tunneling of an electron from the silicon CB is possible. Upon capture, the defect is assumed to undergo a transition to a different and highly excited vibronic state. The corresponding excess energy is dissipated by a multi-phonon emission process, relaxing the system to a lower energy state. The new state has a different barrier

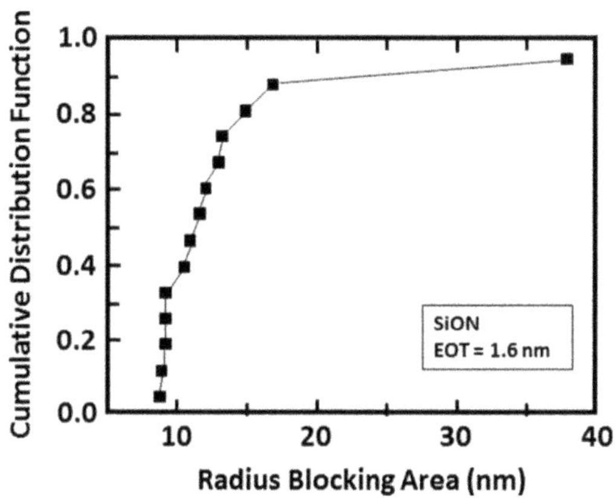

Figure 2.29. Cumulative distribution function of the radius of the blocking area estimated by evaluating the amplitude of RTS in the gate current of nMOSFETs with a SiON dielectric (EOT of 1.6 nm) and polysilicon gate. (Reproduced with permission from Crupi *et al* [79]. Copyright 2009 AIP Publishing.)

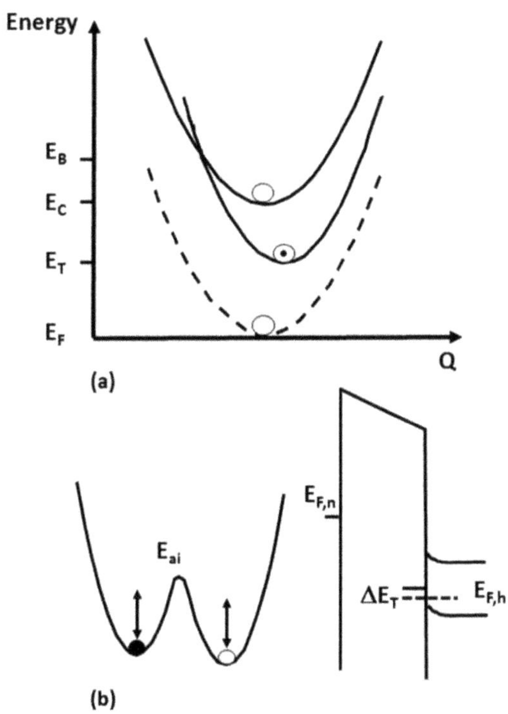

Figure 2.30. Two models to explain TLF dynamics. (*a*) Multi-phonon relaxation model showing a plot versus a normal coordinate Q of the total energy (electronic + vibronic) of an empty trap state with a free electron at the Fermi level (E_F), an empty trap state with a free electron near the bottom of the CB (E_C) and a filled trap state (E_T). The barrier for electron capture is $\Delta E_{CB} = E_B - E_C$, while the barrier for emission is $E_B - E_T$. (*b*) Metastable defect configuration model showing the energy barriers $E_{a,i}$ for defect reconfiguration and the corresponding defect electron energy levels represented by the full and broken curves. ΔE_T represents the shift in defect state energy level upon reconfiguration, with $E_{a,i} > \Delta E_T$. (Reproduced with permission from Farmer and Buhrman [67]. Copyright 1989 IOP Publishing.)

$E_B + (E_C - E_T)$ to overcome before the defect can make a transition back to the original vibronic state after electron tunnel emission back to the CB.

The alternative model in figure 2.30(*b*) considers that the change in the charge state of the defect is a consequence of a thermally activated change in defect configuration, rather than being the origin of it. Defect fluctuations between different (meta-)stable states, separated by an activation barrier are a well-established phenomenon in amorphous systems, leading to the familiar situation of tunneling transitions between two-level states at low temperatures (see also section 2.1.4 and figure 2.20) and activated transitions at higher temperatures. In this case, the oxide-trap state energy is shifted by an amount ΔE_T by the transition to another defect state, which must be sufficiently large to change the occupation probability of the state (i.e. $>kT$). However, this can still be a small fraction of the defect activation energy $E_{a,i}$ if the fluctuation attempt rates are set by the defect atomic vibrational frequency ($\sim 10^{12}$ to 10^{13} s^{-1}) and not by electronic tunneling rates. In this model, the local potential seen by the defect determines the activation energies and fluctuation attempt rates, not the electron energy levels and tunneling capture cross section [67].

Once structural defect reconfigurations are recognized as a source of discrete fluctuations, the necessity that a TLF involves the capture and emission of a single electronic charge disappears. A reconfiguration of a strong, perhaps extended bistable defect could affect maybe more than one electronic state, particularly if such states occur in clusters or chains (silicon surface steps).

Along the same lines, multi-level fluctuations, as depicted in figure 2.31, can be treated as interacting elementary single-electron TLFs which are either attractive or repulsive [71] (see for example also the work of Amarasinghe and Çelik-Butler [23]). A second approach, along the lines of the multi-stable defect model, treats them as thermally activated transitions between different states of an entity (a defect). The multiple levels are then simply a consequence of the different states accessible to the trap and the detailed character of the fluctuations is set by the pathways open to the trap when it is in any particular state [67]. Localized defect excitation by inelastic electron-defect scattering has a significant effect on defect fluctuations between metastable configurations (see also section 2.1.4 for the coupling to an external energy reservoir of a TSS). This gives rise to non-linear $1/T$ and gate voltage dependences of the time constants, indicating that the effective defect temperature is higher than the lattice temperature through excitation of defect vibrational motion [67].

In addition, the relative amplitude of single-point-charge switching on MOS diode conductance has been estimated to be of the order of $<5 \times 10^{-5}$ [67], which is much smaller than the observed amplitudes. If barrier inhomogeneities play a role, a strategically placed trap may have a higher impact. However, both the bias and temperature dependence of TLFs in MOS diodes are more in accord with the defect fluctuation model, whereby not one charge is trapped, but a charge which depends on temperature and V_{GS}. The conductance fluctuations are almost certainly the result of abrupt changes in the amount of charge locally trapped in the oxide; the change does not necessarily amount to a unit charge [67]. The discrete nature of

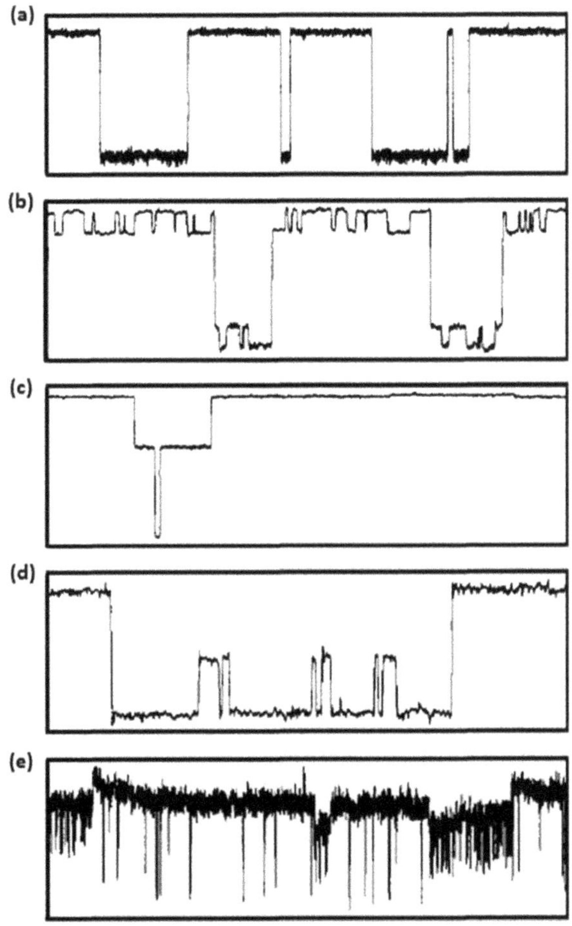

Figure 2.31. Examples of different kinds of multi-level fluctuations observed in 1 μm^2 tunnel diodes (unstressed): (*a*) a single TLF, (*b*) two apparently independent TLFs, (*c*) a three-level fluctuation in which transitions between the highest and lowest conductance levels always occur via an intermediate level, (*d*) a three-level fluctuation in which transitions occur only from the lower level and (*e*) a four-level fluctuation in which the duty cycle and amplitude of one TLF seem to be modulated by the transitions of another TLF. (Reproduced with permission from Farmer and Buhrman [67]. Copyright 1989 IOP Publishing.)

the fluctuations arises instead from the activated transitions of an oxide defect, either localized or extended, between a few accessible metastable configurations, each of which has associated with it a different net trapped charge whose amplitude can vary with temperature and gate bias [67].

An intermediate approach to model a TLF in a silicon MOS diode has been proposed by Andersson *et al* [80], based on trap-assisted electron tunneling via a single trap. The trap not only acts as a single point charge but can function as an intermediate state for electrons that tunnel from the metal to the silicon. The model uses an atomic relaxation effect to account for the thermally activated capture and emission time constants found experimentally [80]. In this interpretation, τ_c is the

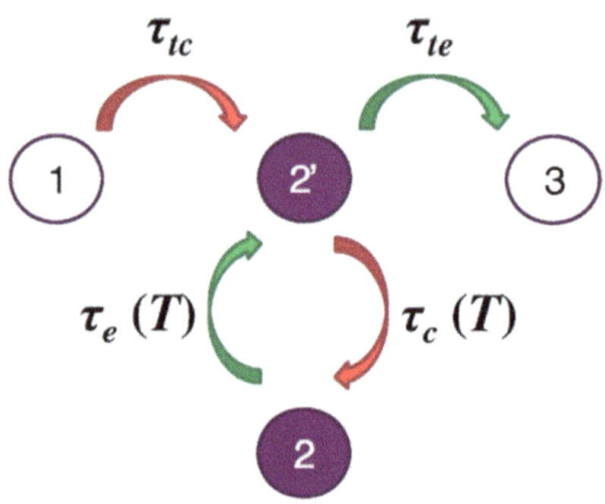

Figure 2.32. State diagram describing the I_G-RTS process. A carrier tunnels from the injecting electrode 1 into a metastable state 2' and either continues to the opposite electrode 3, contributing to ΔI_G or becomes captured in state 2, disabling the conducting path. (Reproduced with permission from Kaczer *et al* [73]. Copyright 2012 Elsevier.)

time constant for electron capture from the metal to the ground state of the trap, and the mean time in the low current state, τ_e, is the time constant for electron emission from the same ground state to the silicon.

A recent critical assessment of I_G RTS in small-area pMOSFETs with equivalent oxide thickness (EOT) = 2.3 nm SiON gate dielectric has concluded that the large amplitudes often found cannot be explained by a local blocking of the gate current by an enhancement of the tunneling barrier [73]. The same argument holds for a model where Coulomb repulsion of the supply carriers in the injecting electrode takes place. A more likely process for the RTS fluctuations in the gate current considers a bistable oxide trap, as depicted in figure 2.32, whereby conduction happens through the trap in a metastable state 2' (by trap-assisted tunneling (TAT)) when it is not occupied by an electron (nFET). This contributes ΔI_G to the total gate current ($I_G = \Delta I_G + I_{DT}$ with I_{DT} the background direct-tunneling current). Such a process would be most efficient for a trap close to the center of the oxide, readily explaining why there are no traps generating both large ΔI_G and ΔI_D in the studied MOSFETs [73]. The similar gate voltage dependence of ΔI_G and I_G can be understood along the same lines. Low conduction (low I_G) occurs when the carrier becomes trapped in the ground state 2 in figure 2.32, which is separated from the tunneling state 2' by an activation barrier E_B. In this picture, the temperature independence of the ΔI_G component is associated with the very fast time constants describing tunneling into and out of the trap. The strong T dependence of the RTS time constants is then due to the thermal barriers between the metastable and stable charge states of the trap. This mechanism is similar to the one proposed earlier for silicon MOS tunneling diodes [80]. It is also compatible with recent insights into the nature of oxide trap originating RTS and BTI in scaled transistors (chapters 3 and 5).

In conclusion, it is clear that extension of the models for channel RTS to the case of gate-current TLFs should be handled with caution. There is ample evidence that the character of RTSs in the gate current I_G may be quite different and the same traps are not necessarily sensed by the channel. This is, in particular, the case after electric stressing of the gate dielectric, as will be seen in chapter 5.

2.4 RTS in the junction leakage current of a MOSFET

As will be seen in chapter 6 on memories, the phenomenon of variable junction leakage (VJL) current can cause a switching in the retention time of DRAMs. It has been assumed that this phenomenon is connected with RTN in the GIDL current [81]. The latter is dominated by interface-state-assisted TAT. There are several possible mechanisms that can cause such VJL [33, 81]:

- Structural fluctuation of a complex defect between two configurations.
- Charge-state fluctuation.
- VJL by a modulation of the generation–recombination (GR) current due to a change in the charge state or occupancy of a trap nearby.

Based on a dedicated measurement method of VJL, it has been shown that a hole trap in the gate oxide could explain the observations and in particular why activation energies of the up and down times larger than $E_G/2$ can be found [81]. The VJL is measured by connecting many transistors in parallel so that the sum of the (small) junction leakage currents is measured. As VJL occurs only in a small percentage of devices, a large number of them have to be measured. The basic mechanism behind VJL is sketched in figures 2.33 and 2.34.

The drain–bulk junction leakage current is generated in the transistor off-state and is observed as a current between the drain and substrate. Charge capture and emission in the gate oxide by a nearby trap acts as the origin of the two-state fluctuation of this leakage [81]. According to figure 2.34, a higher leakage current is found when the trap becomes positively charged upon hole capture, since a higher electric field exists in the depletion region. In order to emphasize the hole trapping during the measurements, applying a gate voltage pulse into accumulation has been proposed, which enhances the probability for hole capture by the gate oxide traps [81]. In addition, an increase of the density of interface states by HC stress will increase the likelihood of VJL.

VJL can be modeled by considering the transitions in figure 2.35 [81]. In the case of hole emission, electrons are thermally excited from the valence band to the trap level and subsequently tunnel to the trap. The corresponding transition rate is given by $R_e = R_{thr} \times P_{ctun}$ [81], with

$$R_{thr} = v_{th} \times \sigma_p \times N_v \exp\left(-\frac{E_T - E_V}{kT}\right) \tag{2.34}$$

and

$$P_{ctun} = \exp\left(-2d\sqrt{2\pi m_e^*(qV_{oxe})/\hbar}\right) \tag{2.35}$$

with V_{oxe} the potential barrier from silicon to SiO_2 for an electron and d the tunneling distance.

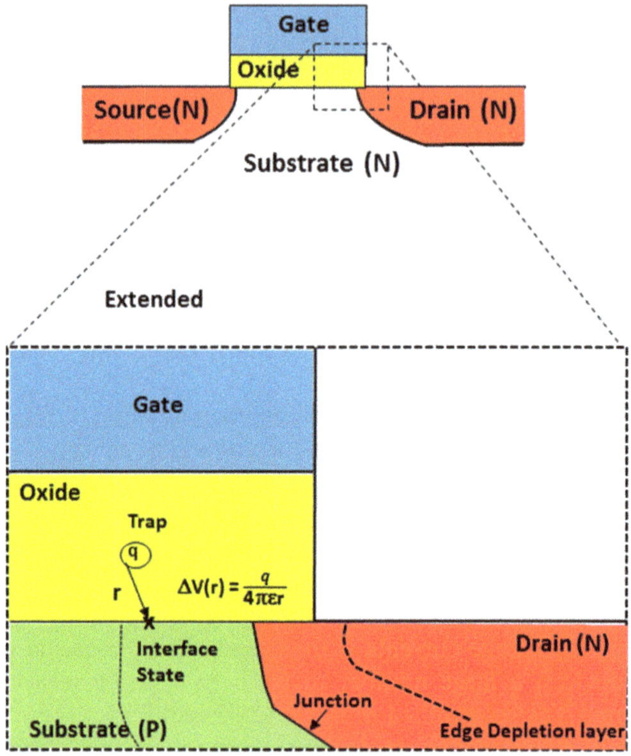

Figure 2.33. Potential modulation (ΔV) at interface states caused by a charge capture at a trap site located in the gate oxide. (After Mori *et al* [81].)

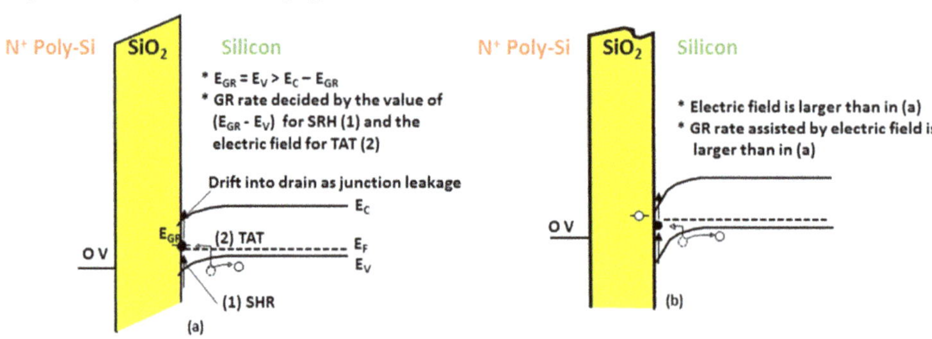

Figure 2.34. The GR rate at interface states is influenced by traps in the gate oxide near the Si/SiO$_2$ interface. The case of a hole trap in SiO$_2$ is shown as an example: (*a*) No carrier trapped in the SiO$_2$. (*b*) A hole trapped in the SiO$_2$. The GR rate at the interface states near the trap is higher than in case (*a*). (After Mori *et al* [81].)

For the hole capture, one considers electron–hole pair formation across the gap, whereby the hole tunnels to the trap. This gives a transition rate $R_{\rm h} = R_{\rm gen}P_{\rm htun}$ [85], corresponding to

$$R_{\rm gen} = v_{\rm th} \times \sigma_{\rm gen} \times N_v \exp\left(-\frac{E_{\rm GR} - E_{\rm V}}{kT}\right) \tag{2.36}$$

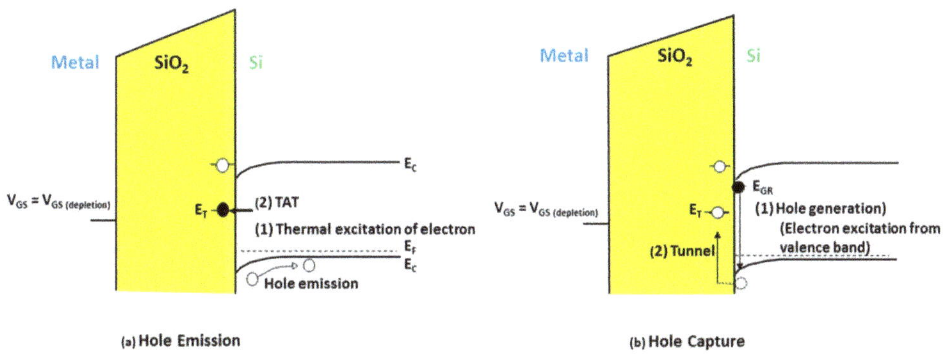

Figure 2.35. Models for (*a*) hole emission and (*b*) hole capture. (After Mori *et al* [81].)

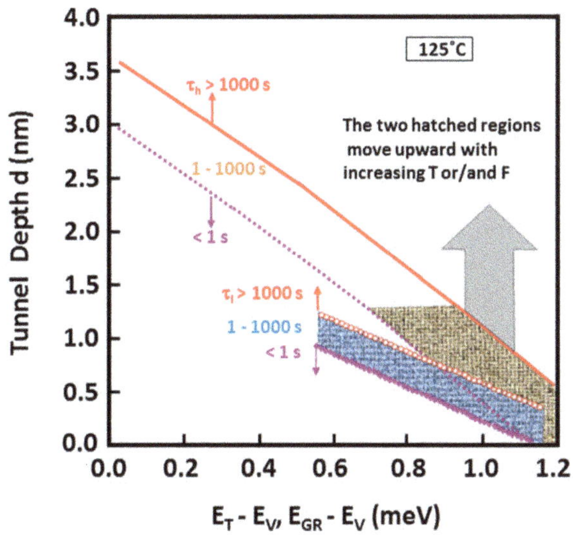

Figure 2.36. Relationship between d and $(E_T - E_V)$ for $\langle\tau_h\rangle$ and between d and $(E_{GR} - E_V)$ for $\langle\tau_l\rangle$. The region hatched with diagonal lines indicates the condition in which both time constants can take up values between 1 s and 1000 s under a constant d condition (d is constant in both capture and emission processes). (After Mori *et al* [81].)

and

$$P_{htun} = \exp\left(-2d\sqrt{2\pi m_h^*(qV_{oxh})/\hbar}\right) \tag{2.37}$$

In this way, activation energies $>E_G/2$ can be obtained (E_{GR}), depending on the electron–hole generation mechanism involved (band-to-band, via a trap state, etc) [81]. The obtained tunneling depths versus the activation energies are depicted in figure 2.36. From this, one can derive that a tenfold increase in VJL is estimated when a trap at a depth $r = 1.8$ nm becomes occupied by a hole [81].

Direct evidence that interface states are involved in the VJL phenomenon has been derived from the study of the occurrence rate of VJL in wafers with different

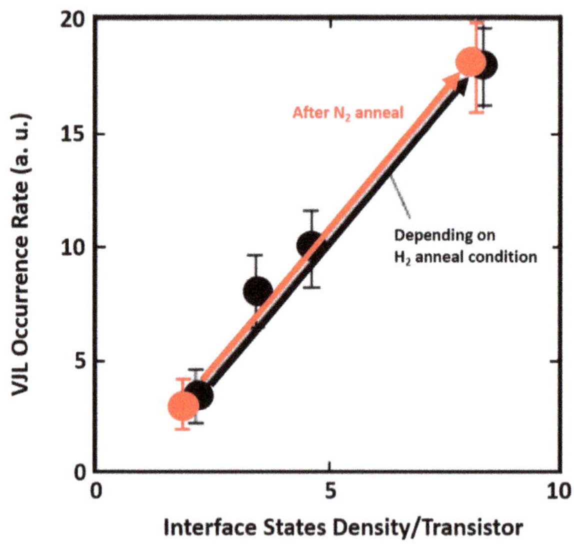

Figure 2.37. Relationship between occurrence rate of VJL and density of interface states per transistor. The error bar means a 95% confidence interval. (After Mori *et al* [82].)

D_{it} [82]. These have been prepared by either annealing at 400 °C in forming gas (FGA) or in nitrogen. While the former decreases the D_{it}, in the latter case an increase is found. According to figure 2.37, there exists a nice linear correlation between the occurrence rate and D_{it}, in strong support of the oxide-trap modulation of the surface generation current [86]. In fact, this mechanism is rather similar to the RTS in the GIDL current of a MOSFET [37], mentioned earlier.

2.5 Multiple and complex RTS

The previous case of RTS in the substrate current was already an example of mutual interaction between an oxide trap and surface generation through interface states. This implies that in devices with several oxide traps present, complex RTS can be observed [18, 83, 84]. A detailed analysis of 0.06 μm^2 medium doped drain nMOSFETs in strong inversion [18] has revealed the presence of three acceptor traps observable at the same time in a specific V_{GS} range. It was even found that the energy level of one trap was upward shifted when another one was occupied. This can be understood considering that two nearby repulsive centers will influence each other when charged [18].

In other cases, anomalous two-level RTSs, as in figure 2.38 [83], have been observed. During a certain period random current switching is observed, and between t_2 and t_3 no trapping occurs. This behavior has been explained in terms of a trap which occurs in different metastable states. In this case, the interpretation of the switching times and activation energies also becomes more complicated [83]. These anomalous signals bear a strong resemblance to the RTSs which have been found in the tunneling current of an MOS diode, as was shown in a previous section. They are also believed to give rise to non-Gaussian noise. In addition, three-level

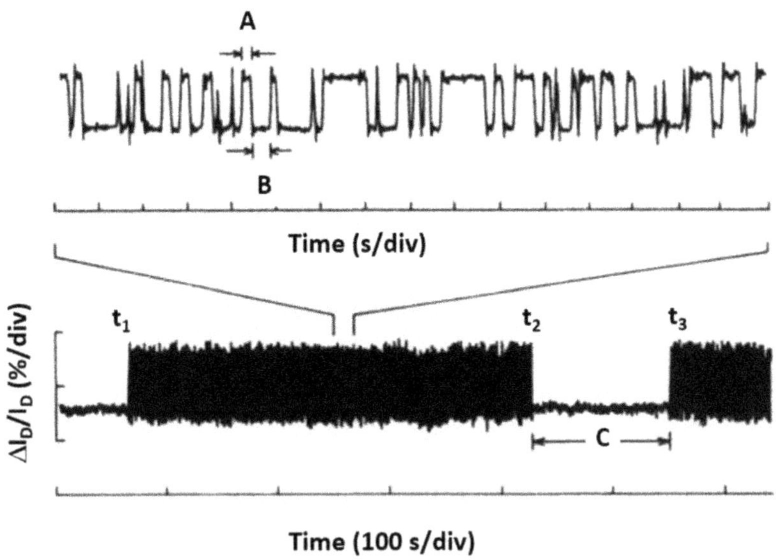

Figure 2.38. Fluctuations in drain current versus time showing a rapidly switching RTS modulated by an envelope of the same amplitude. The upper trace is an expansion of the lower trace. $V_{GS} = 1$ V, $V_{DS} = 10$ mV; $I_D = 10.3$ nA, $T = 293$ K. (Reproduced with permission from Uren *et al* [83]. Copyright 1988 the American Physical Society.)

RTSs from a defect showing one- and two-electron capture have also been reported [83]. In this context, a basic introduction to defect bistability can be found in [85].

Several studies have pointed out that due to Coulomb repulsion, the occupation statistics of interacting or coupled oxide traps changes [86–88], leading to so-called state-loss in complex RTS. Normally, when n traps are present one expects 2^n current levels, which may appear in, for example, a time-lag plot (TLP) [88]. However, it has been observed that for interacting traps not all states may be retrieved [87, 88]. A dedicated model has been established in order to address the consequences on a circuit level with respect to the dynamic variability [87]. According to the results of figure 2.39 [86], the coupling should decrease for increasing gate voltage (or current). The coupling effect of a filled trap stems from the image charge in the substrate, given by $q' = q\,(\varepsilon_{ox} - \varepsilon_{sem})/(\varepsilon_{ox} + \varepsilon_{sem})$ with ε_{ox} and ε_{sem} the dielectric constant of the oxide and the semiconductor, respectively. Since $\varepsilon_{ox} > \varepsilon_{sem}$ for the high-κ gate dielectrics studied in this case, the polarity of the image charge is the same as that of the trapped charge. This negative charge on the trap and in the substrate will apply a negative potential on a neighboring trap, raising its energy level by an amount given by the Coulomb energy E_{Coul}. This implies that the capture barrier will become higher. The τ_c of a nearby oxide trap can be expressed as

$$\frac{1}{\tau_c} = v_{thn}\sigma_n n \exp\left(\frac{E_B + E_{Coul}}{kT}\right) \tag{2.38}$$

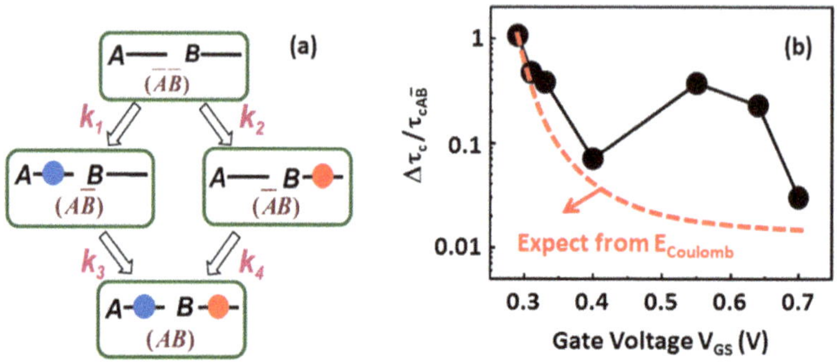

Figure 2.39. (*a*) The transition process of the four states of the two traps (– empty; ● filled). (*b*) V_{GS} dependence of the extracted trap coupling factor compared to the expectations from E_{Coul} (dashed line). A non-monotonous trend is observed. (Reproduced with permission from Ren *et al* [86]. Copyright 2014 AIP Publishing.)

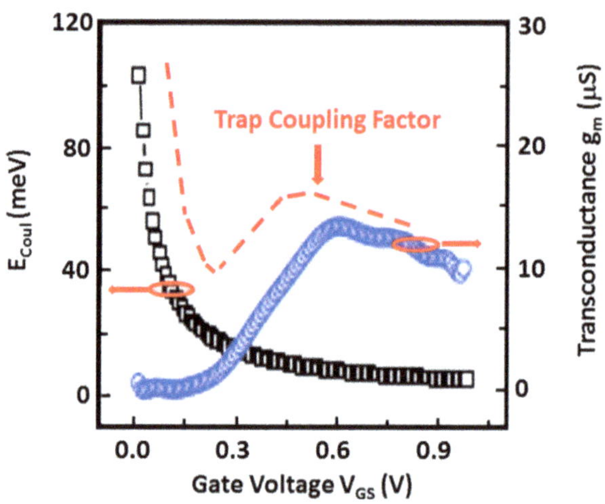

Figure 2.40. The trap coupling factor effect caused by two competing mechanisms, namely, the Coulomb repulsion effect (indicated by E_{Coul}) and the local carrier density perturbation effect (indicated by g_m). (Reproduced with permission from Ren *et al* [86]. Copyright 2014 AIP Publishing).

with

$$E_{Coul} \approx \frac{q^2}{4\pi\varepsilon_{ox}r_0}\left\{\frac{1+[qt_{inv}C_{ox}(V_{GS}-V_T)}{4kT\varepsilon_{ox}WL}\right\}^{-1} \tag{2.39}$$

In equation (2.39) r_0 is the distance between the two traps. The trap coupling factor can be expressed as [86]

$$\frac{\Delta\tau_c}{\tau_{cA\overline{B}}} = \frac{\tau_{cAB}-\tau_{cA\overline{B}}}{\tau_{cA\overline{B}}} \tag{2.40}$$

with $\tau_{cA\overline{B}}$ the capture time constant of trap A with trap B empty ($\overline{B}$) and τ_{cAB} is the capture time constant of trap A with trap B occupied.

It is expected that at higher V_{GS} (nMOSFET) the increased carrier density will screen the image charge so that E_{Coul} and the trap coupling will decrease. However, according to figure 2.40, the decrease of the coupling factor is non-monotonous, in contrast to the Coulomb repulsion effect [86]. Additional elements can be taken into account, such as the impact of the local carrier density in nanoscale devices, which is determined by the local potential landscape induced by the random dopants in the channel. According to atomistic simulations, more percolation paths (PPs) open at intermediate gate voltage, which merge at higher V_{GS}, so that the classical Coulomb repulsion behavior should be retrieved. Overall, the trap coupling factor follows the trend of the transconductance with gate voltage, as shown in figure 2.40 [86].

References

[1] Kirton M J, Uren M J, Collins S, Schulz M, Karmann A and Scheffer K 1989 Individual defects at the Si:SiO_2 interface *Semicond. Sci. Technol.* **4** 1116–26

[2] Ralls K S, Skocpol W J, Jackel L D, Howard R E, Fetter L A, Epworth R W and Tennant D M 1984 Discrete resistance switching in submicrometer inversion layers: individual interface traps and low-frequency (1/*f*?) noise *Phys. Rev. Lett.* **52** 228–31

[3] Kirton M J and Uren M J 1986 Capture and emission kinetics of individual Si:SiO_2 interface states *Appl. Phys. Lett.* **48** 1270–2

[4] Amarasinghe N V, Çelik-Butler Z and Vasina P 2000 Characterization of oxide traps in 0.15 μm^2 MOSFETs using random telegraph signals *Microelectron. Reliab.* **40** 1875–81

[5] Mueller H H, Wörle D and Schulz M 1994 Evaluation of the Coulomb energy for single-electron interface trapping in sub-μm metal–oxide–semiconductor field-effect transistors *J. Appl. Phys.* **75** 2970–9

[6] Cobden D H and Uren M J 1993 Random telegraph signals from liquid helium to room temperature *Microelectron. Eng.* **22** 163–70

[7] Schulz M and Karmann A 1991 Individual, attractive defect centers in the Si–SiO_2 interface of μm-sized MOSFETs *Appl. Phys.* A **52** 104–11

[8] Shi Z, Miéville J-P and Dutoit M 1994 Random telegraph signals in deep submicron n-MOSFETs *IEEE Trans. Electron Devices* **41** 1161–8

[9] Shi Z M, Miéville J P and Dutoit M 1992 Effect of electron heating on RTS in deep submicron n-MOSFETs *Microelectron. Eng.* **19** 751–4

[10] Wirth G, Hilleringmann U and Goser K 2000 Single electron switching events in ultrashort bulk MOSFETs *Proc. 15th SBMicro—Int. Conf. on Microelectron. and Packaging* ed N L A Braga, S Bampi and K Shenai Sept. 18–23, 2000, (Manaus, Brazil) pp 108–13

[11] Karwath A and Schulz M 1988 Deep level transient spectroscopy on single, isolated interface traps in field-effect transistors *Appl. Phys. Lett.* **52** 634–6

[12] Schulz M and Karmann A 1991 Single, individual traps in MOSFETs *Phys. Scr.* **T35** 273–80

[13] Grasser T, Reisinger H, Wagner P-J, Schanovsky F, Goes W and Kaczer B 2010 The time dependent defect spectroscopy (TDDS) for the characterization of the bias temperature instability *Proc. Int. Reliab. Phys. Symp.—IRPS10* (New York: IEEE) pp 16–25

[14] Grasser T, Reisinger H, Wagner P-J and Kaczer B 2010 Time-dependent defect spectroscopy for characterization of border traps in metal–oxide–semiconductor transistors *Phys. Rev.* B **82** 245318

[15] Kirton M J and Uren M J 1989 Noise in solid-state microstructures: a new perspective on individual defects, interface states and low-frequency (1/*f*) noise *Adv. Phys.* **38** 367–468

[16] Veksler D, Bersuker G, Rumyantsev S, Shur M, Park H, Young C, Lim K Y, Taylor W and Jammy R 2010 Understanding noise measurements in MOSFETs: the role of traps structural relaxation *Proc. Int. Reliability Physics Symp.—IRPS10* (New York: IEEE) pp 73–9

[17] Cobden D H, Uren M J and Kirton M J 1990 Entropy measurements on slow Si/SiO_2 interface states *Appl. Phys. Lett.* **56** 1245–7

[18] Amarasinghe N V and Çelik-Butler Z 2000 Complex random telegraph signals in 0.06 μm^2 MDD n-MOSFETs *Solid-State Electron.* **44** 1013–9

[19] Schulz M, Pappas A and Vennemann J 1993 Single-electron trapping and telegraph noise in μm-sized MOSFETs *The Physics and Chemistry of SiO_2 and the Si:SiO_2 Interface* vol 2 ed C R Helms and B E Deal (New York: Plenum) pp 383–92

[20] Çelik-Butler Z, Vasina P and Amarasinghe N V 2000 A method for locating the position of oxide traps responsible for random telegraph signals in submicron MOSFETs *IEEE Trans. Electron Devices* **47** 646–8

[21] Tsai M-H, Muto H and Ma T P 1992 Random telegraph signals arising from fast interface states in metal–SiO_2–Si transistors *Appl. Phys. Lett.* **61** 1691–3

[22] Kandiah K, Deighton M O and Whiting F B 1989 A physical model for random telegraph signal currents in semiconductor devices *J. Appl. Phys.* **66** 937–48

[23] Nagumo T, Takeuchi K, Hase T and Hayashi Y 2010 Statistical characterization of trap position, energy, amplitude and time constants by RTN measurement of multiple individual traps *Tech. Dig. Int. Electron Devices Meeting—IEDM10* (New York: IEEE) pp 628–31

[24] Saks N S, Groeseneken G and De Wolf I 1996 Characterization of individual interface traps with charge pumping *Appl. Phys. Lett.* **68** 1383–5

[25] Wu J W, You J W, Ma H C, Cheng C C, Chang C S, Huang G W and Wang T 2004 Valence-band tunneling induced low frequency noise in ultrathin oxide (15 Å) n-type metal–oxide–semiconductor field effect transistors *Appl. Phys. Lett.* **85** 5076–7

[26] Wu J-W, You J-W, Ma H-C, Cheng C-C, Hsu C-F, Chang C-S, Huang G-W and Wang T 2005 Excess low-frequency noise in ultrathin oxide n-MOSFETs arising from valence-band electron tunneling *IEEE Trans. Electron Devices* **52** 2061–6

[27] Wu J W, You J W, Ma H C, Cheng C C, Hsu C F, Huang G W, Chang C S and Wang T 2005 Low frequency noise degradation in ultra-thin oxide (15 Å) analog n-MOSFETs resulting from valence-band tunneling *Proc. 43rd Annual Int. Reliability Physics Symp.—IRPS05* (New York: IEEE) pp 260–4

[28] Simoen E and Claeys C 1994 Random telegraph signals in silicon-on-insulator metal–oxide–semiconductor transistors *J. Appl. Phys.* **75** 3647

[29] Simoen E and Claeys C 1996 The low-frequency noise behaviour of silicon-on-Insulator technologies *Solid-State Electron.* **39** 949–60

[30] Simoen E, Magnusson U, Vermeiren J and Claeys C 1993 Back-gate induced random telegraph signal noise in fully-depleted silicon-on-insulator nMOSFETs *Solid-State Electron.* **36** 1593–6

[31] Silva H and Tiwari S 2006 Random telegraph signals in nanoscale back-side charge trapping memories *Appl. Phys. Lett.* **88** 102105

[32] Lee J-W, Shin H and Lee J-H 2010 Characterization of random telegraph noise in gate induced drain leakage current of n- and p-type metal–oxide–semiconductor field-effect transistors *Appl. Phys. Lett.* **96** 043502

[33] Mori Y, Takeda K and Yamada R 2010 Random telegraph noise of junction leakage current in submicron devices *J. Appl. Phys.* **107** 014509

[34] Lukyanchikova N B, Petrichuk M V, Garbar N P, Simoen E and Claeys C 1998 Impact of the free electron distribution on the random telegraph signal capture kinetics in submicron n-metal–oxide–semiconductor field-effect transistors *Appl. Phys. Lett.* **73** 2444–6

[35] Schulz M 1993 Coulomb energy of traps in semiconductor space-charge regions *J. Appl. Phys.* **74** 2649–57

[36] Campbell J P, Qin J, Cheung K P, Yu L C, Suehle J S, Oates A and Sheng K 2009 Random telegraph noise in highly scaled nMOSFETs *Proc. 47th Annual Int. Reliability Physics Symp.—IRPS09* (New York: IEEE) pp 382–8

[37] Ferraton S, Zimmermann J, Montès L, Ghibaudo G, Brini J, Gurgul J and Chroboczek J A 2004 Direct extraction of McWhorter's constant from LFN spectra of MOSFETs with planar layers of Si nano-crystals embedded in gate SiO_2 *Proc. SPIE* **5470** 560–5

[38] Shi Y, Bu H M, Yuan X L, Gu S L, Shen B, Han P, Zhang R and Zheng Y D 2001 Switching kinetics of interface states in deep submicrometre SOI n-MOSFETs *Semicond. Sci. Technol.* **16** 21–5

[39] Scofield J H, Borland N and Fleetwood D M 2000 Temperature-independent switching rates for a random telegraph signal in a silicon metal–oxide–semiconductor field-effect transistor at low temperatures *Appl. Phys. Lett.* **76** 3248–50

[40] Vuillaume D, Bourgoin J C and Lannoo M 1986 Oxide traps in Si–SiO_2 structures characterized by tunnel emission with deep-level transient spectroscopy *Phys. Rev.* B **34** 1171–83

[41] Gusmeroli R, Monzio Compagnoni C, Riva A, Spinelli A S, Lacaita A L, Bonanomi M and Visconti A 2006 Defects spectroscopy in SiO_2 by statistical random telegraph noise analysis *Tech. Dig. Int. Electron Devices Meeting—IEDM06* (New York: IEEE) pp 483–6

[42] Monzio Compagnoni C, Gusmeroli R, Spinelli A S and Visconti A 2008 RTN V_T instability from the stationary trap-filling condition: an analytical spectroscopic investigation *IEEE Trans. Electron Devices* **55** 655–61

[43] Cobden D H, Uren M J and Pepper M 1993 Dissipative tunneling in two-state systems at the Si/SiO_2 interface *Phys. Rev. Lett.* **71** 4230–3

[44] Cobden D H and Muzykantskii B A 1995 Finite-temperature Fermi-edge singularity in tunneling studied using random telegraph signals *Phys. Rev. Lett.* **75** 4274–7

[45] Uren M J, Day D J and Kirton M J 1985 $1/f$ and random telegraph noise in silicon metal-oxide-semiconductor field-effect transistors *Appl. Phys. Lett.* **47** 1195–7

[46] Simoen E, Dierickx B, De Canne B, Thoma F and Claeys C 1994 On the gate- and drain voltage dependence of the RTS amplitude in submicron MOSTs *Appl. Phys.* A **58** 353–8

[47] Mueller H H, Schirl U and Schulz M 1996 Charge carrier fluctuations in sub-μm MOSFETs *Proc. 3rd Int. Symp. on the Physics and Chemistry of SiO_2 and the Si–SiO_2 Interface* ; Massoud H Z, Poindexter E H and Helms C R *ECS Proc.* Vol 96 pp 568–79

[48] Mueller H H and Schulz M 1996 Conductance modulation of submicrometer metal–oxide–semiconductor field-effect transistors by single-electron trapping *J. Appl. Phys.* **79** 4178–86

[49] Vandamme L K J, Sodini D and Gingl Z 1998 On the anomalous behavior of the relative amplitude of RTS noise *Solid-State Electron.* **42** 901–5

[50] Hung K K, Ko P K, Hu C and Peng Y C 1990 Random telegraph noise of deep-submicrometer MOSFETs *IEEE Electron Device Lett.* **11** 90–2

[51] Godoy A, Gámiz F, Palma A, Jiménez-Tejada J A and Carceller J E 1997 Random telegraph signal amplitude in submicron n-channel metal oxide semiconductor field-effect transistors *Appl. Phys. Lett.* **70** 2153–5

[52] Mueller H H and Schulz M 1997 Statistical evaluation of random telegraph signal amplitudes in sub-μm MOSFETs *Microelectron. Reliab.* **36** 223–6
[53] Mueller H H and Schulz M 1998 Random telegraph signal: an atomic probe of the local current in field-effect transistors *J. Appl. Phys.* **83** 1734–41
[54] Ishikuro H, Saraya T, Hiramoto T and Ikoma T 1996 Extremely large amplitude random telegraph signals in a very narrow split-gate MOSFET at low temperatures *Jpn. J. Appl. Phys.* **35** 858–60
[55] Fantini P, Ghetti A, Marinoni A, Ghidini G, Visconti A and Marmiroli A 2007 Giant random telegraph signals in nanoscale floating-gate devices *IEEE Trans. Electron Devices* **28** 1114–6
[56] Prati E, Fanciulli M, Ferrari G and Sampietro M 2008 Giant random telegraph signal generated by single charge trapping in submicron n-metal–oxide–semiconductor field-effect transistors *J. Appl. Phys.* **103** 123707
[57] Restle P and Gnudi A 1990 Internal probing of submicron FETs and photoemission using individual oxide traps *IBM J. Res. Dev.* **34** 227–42
[58] Restle P 1988 Individual oxide traps as probes into submicron devices *Appl. Phys. Lett.* **53** 1862–4
[59] Roux dit Buisson O, Ghibaudo G and Brini J 1992 Model for drain current RTS amplitude in small-area MOS transistors *Solid-State Electron.* **35** 1273–6
[60] Tsai M-H, Ma T P and Hook T B 1994 Channel length dependence of random telegraph signal in sub-micron MOSFETs *IEEE Electron Device Lett.* **15** 504–6
[61] Tsai M-H and Ma T P 1994 The impact of device scaling on the current fluctuations in MOSFETs *IEEE Trans. Electron Devices* **41** 2061–7
[62] Bu H M, Shi Y, Yuan X L, Wu J, Gu S L, Zheng Y D, Majima H, Ishikuro H and Hiramoto T 2000 Random telegraph signals and low-frequency noise in n-metal–oxide–semiconductor field-effect transistors with ultranarrow channels *Appl. Phys. Lett.* **76** 3259–61
[63] Bu H M, Shi Y, Yuan X L, Zheng Y D, Gu S H, Majima H, Ishikuro H and Hiramoto T 2000 Impact of device scaling on the low-frequency noise in n-MOSFETs *Appl. Phys.* A **71** 133–6
[64] Zhang P, Zhuang Y Q, Ma Z F, Bao L, Du L and Bao J L 2008 The impact of gate-image charge on RTS amplitudes in ultra-thin gate oxide n-MOSFETs *Semicond. Sci. Technol.* **23** 125037–1
[65] Welland M E and Koch R H 1986 Spatial location of electron trapping defects on silicon by scanning tunneling microscopy *Appl. Phys. Lett.* **48** 724–6
[66] Farmer K R, Rogers C T and Buhrman R A 1987 Localized-state interactions in metal-oxide-semiconductor tunnel diodes *Phys. Rev. Lett.* **58** 2255–8
[67] Farmer K R and Buhrman R A 1989 Defect dynamics and wear-out in thin silicon oxides *Semicond. Sci. Technol.* **4** 1084–108
[68] Avellan A, Krautschneider W and Schwantes S 2001 Observation and modeling of random telegraph signals in the gate and drain current of tunneling metal–oxide–semiconductor field-effect transistors *Appl. Phys. Lett.* **78** 2790–2
[69] Avellan A, Krautschneider W and Sell B 2001 Softbreakdown and RTS and drain currents of MOS transistors with ultrathin oxides *Proc. of the 16th Int. Conf. on Noise in Physical Systems and 1/f Fluctuations—ICNF 2001* ed G Bosman (Singapore: World Scientific) pp 375–8

[70] Avellan A, Schroeder D and Krautschneider W 2003 Modeling random telegraph signals in the gate current of metal–oxide–semiconductor field-effect transistors after oxide breakdown *J. Appl. Phys.* **94** 703–8
[71] Martinez F *et al* 2005 Oxide traps characterization of 45 nm MOS transistors by gate current RTS noise measurements *Microelectron. Reliab.* **80** 54–5
[72] Armand J, Martinez F, Valenza M, Rocherau K and Vincent E 2006 Slow oxide trap density profile extraction using gate current low-frequency noise in ultrathin oxide MOSFETs *Microelectron. Eng.* **84** 2382–5
[73] Kaczer B, Toledano-Luque M, Goes W, Grasser T and Groeseneken G 2013 Gate current random telegraph noise and single defect conduction *Microelectron. Eng.* **109** 123–5
[74] Chang C M, Chung S S, Hsieh Y S, Cheng L W, Tsai C T, Ma G H, Chien S C and Chun S W 2008 The observation of trapping and detrapping effects in high-*k* gate dielectric MOSFETs by a new gate current random telegraph noise (I_G-RTN) approach *Tech. Dig. Int. Electron Devices Meeting—IEDM08* (New York: IEEE)
[75] Lee S, Cho H-J, Son Y, Lee D S and Shin H 2009 Characterization of oxide traps leading to RTN in high-*k* and metal gate MOSFETs *Tech. Dig. Int. Electron Devices Meeting—IEDM09* (New York: IEEE) pp 763–6
[76] Cho H-J, Lee S, Park B-G and Shin H 2010 Extraction of trap energy and location from random telegraph noise in gate leakage current (I_g RTN) of metal–oxide–semiconductor field effect transistor (MOSFET) *Solid-State Electron.* **54** 362–7
[77] Cho H-J, Son Y, Oh B-C, Lee S, Lee J-H, Park B-G and Shin H 2010 Study on time constants of random telegraph noise in gate leakage current through hot-carrier stress test *IEEE Electron Device. Lett.* **31** 1029–31
[78] Magnone P, Crupi F, Giusi G, Pace C, Simoen E, Claeys C, Maji D, Ramgopal Rao V and Srinivasan P 2009 $1/f$ noise in drain and gate current of MOSFETs with high-*k* gate stack *IEEE Trans. Device Mater. Reliab.* **9** 180–9
[79] Crupi F, Giusi G, Iannaccone G, Magnone P, Pace C, Simoen E and Claeys C 2009 Analytical model for the $1/f$ noise in the tunneling current through metal–oxide–semiconductor structures *J. Appl. Phys.* **106** 073710–1
[80] Andersson M O, Xiao Z, Norrman S and Engström O 1990 Model based on trap-assisted tunneling for two-level current fluctuations in submicrometer metal–oxide–silicon diodes *Phys. Rev.* B **41** 9836–42
[81] Mori Y, Yoshimoto H, Takeda K and Yamada R 2012 Mechanism of random telegraph noise in junction leakage current of metal–oxide–semiconductor field-effect transistor *J. Appl. Phys.* **111** 104513/1–9
[82] Mori Y, Shima A, Takeda K and Yamada R 2012 Experimental evidence for involvement of interface states in random telegraph noise in junction leakage current of metal–oxide–semiconductor field-effect transistor *Jpn. J. Appl. Phys.* **51** 104101
[83] Uren M J, Kirton M J and Collins S 1988 Anomalous telegraph noise in small-area silicon metal–oxide–semiconductor field-effect transistors *Phys. Rev.* B **37** 8346–50
[84] Ohata A, Toriumi A, Iwase M and Natori K 1990 Observation of random telegraph signals: anomalous nature of defects at the Si/SiO_2 interface *J. Appl. Phys.* **68** 200–4
[85] Chantre A 1989 Introduction to defect bistability *Appl. Phys.* A **48** 3–9
[86] Ren P, Wang R, Jiang X, Qiu Y, Liu C and Huang R 2014 Experimental study on the oxide trap coupling effect in metal oxide semiconductor field effect transistors with HfO_2 gate dielectrics *Appl. Phys. Lett.* **104** 263512

[87] Zou J *et al* 2014 New understanding of state-loss in complex RTN: statistical experimental study, trap interaction models, and impact on circuits *Tech. Dig. Int. Electron Devices Meeting—IEDM14* (New York: IEEE) pp 832–5

[88] Obara T, Teramoto A, Yonezawa A, Kuroda R, Sugawa S and Ohmi T 2014 Analyzing correlation between multiple traps in RTN characteristics *Proc. Annual Int. Reliability Physics Symp.—IRPS14* (New York: IEEE) pp 1–6

IOP Publishing

Eddy Simoen and Cor Claeys

Chapter 3

RTS modeling, simulation and parameter extraction

This chapter discusses the modeling and simulation of RTS parameters. In particular for the amplitude, one has to rely on numerical device simulations to reach a sufficient level of understanding and agreement with measurement data. In addition, for present-day deeply scaled devices, the inversion-layer quantization has to be considered as well in the modeling of the time constants—especially for the capture time and the Coulomb barrier.

3.1 Time constant modeling and simulation

Palma *et al* [1] were the first to consider a full quantum 2D calculation of the RTS time constants. Later, less elaborate analytical models have been worked out [2–4]. It has been pointed out that inversion-layer quantization effects become important at the 10^{17} cm^{-3} substrate doping level and for $t_{ox} < 5$ nm [4]. One of the main consequences is that the bottom of the CB increases with ΔE_0 above E_C, given in first approximation by

$$\Delta E_0 = (\hbar/2m_l)^{1/3}(9\pi q F_s/8)^{2/3} \tag{3.1}$$

and the charge at the interface becomes zero and peaks at the charge centroid 3/*b*, with

$$b = \left[12qm_l/\hbar^2\varepsilon_{Si}\varepsilon_0\left(Q_B + 11/32Q_{inv}\right)\right]^{1/3} \tag{3.2}$$

In equation (3.2), Q_B and Q_{inv} are the depletion and inversion charge density, respectively, and the surface electric field F_s is given by $(Q_B + Q_{inv})/\varepsilon_{Si}\varepsilon_0$ (ε_0 is the permittivity of the vacuum). Finally, m_l is the longitudinal effective mass of the carriers.

Implementing inversion-layer quantization in the capture and emission time constants yields [4]

$$\tau_e = \frac{\exp[(E_C - E_T + \Delta E_0)/kT]}{\sigma_n(2D)v_{thn}N_C'(2D)} \tag{3.3a}$$

$$\tau_c = \frac{\exp[(E_C - E_F + \Delta E_0)/kT]}{\sigma_n(2D)v_{thn}N_C'(2D)} \tag{3.3b}$$

and

$$N_C'(2D) \approx 2kTm_l b5\pi\hbar^2 \tag{3.3c}$$

is the effective density of states for the 2D quantized inversion-layer case. Note that the capture cross section in equation (3.3) is also modified by quantization. The RTS amplitude given by equation (2.26) will be modified since both the surface carrier density and the mobility are affected by the high vertical electric field. In addition, it is also derived that the contribution of the number fluctuations to the RTS amplitude will increase with increasing (forward) substrate bias. As demonstrated in figure 3.1, the RTS amplitude indeed increases with forward V_{BS} in an n-channel transistor [3]. The same applies for the scattering coefficient in figure 3.2.

As mentioned before, the RTS up and down time constants not only reveal details about the trap character (donor, acceptor) from the gate voltage dependence (or rather the carrier density dependence) and on the energy level and cross section (temperature dependence, Arrhenius plot), but the gate voltage dependence also enables one, in principle, to extract the depth of the trap in the dielectric, while from

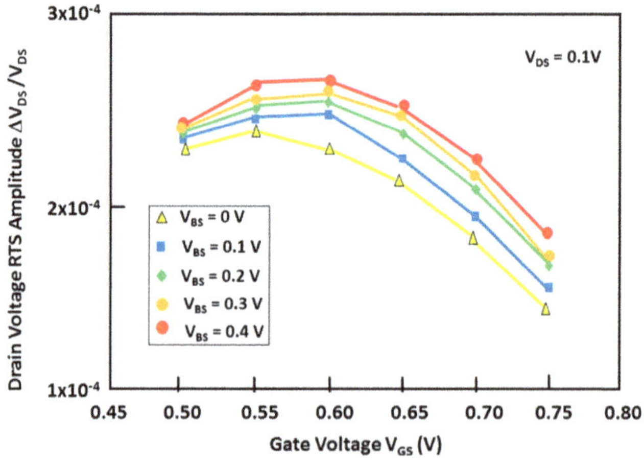

Figure 3.1. RTS amplitude $\Delta V_{DS}/V_{DS}$ at $V_{DS} = 0.1$ V. (Reproduced with permission from Çelik-Butler and Wang [3]. Copyright 2000 Elsevier.)

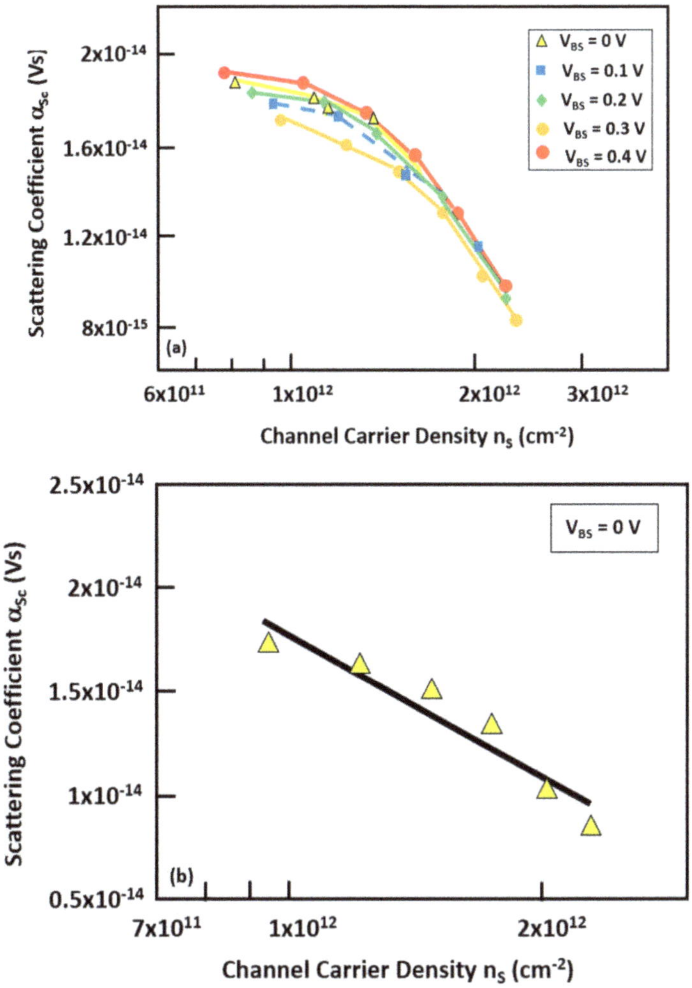

Figure 3.2. (*a*) Scattering coefficient versus channel carrier density obtained from RTS amplitude. (*b*) Scattering coefficient under zero bulk bias, fitted by $\alpha_{sc} = 2.91 \times 10^{-13} - 9.93 \times 10^{-15} \ln(n_s)$ (Vs). (Reproduced with permission from Çelik-Butler and Wang [3]. Copyright 2000 Elsevier.)

the drain voltage dependence in forward and reverse operation the lateral position is extracted.

A correct extraction procedure should proceed according to the following steps [1]:

- The Coulomb energy values are determined as a function of temperature and gate voltage.
- The trap depth and lateral position are derived from τ_c/τ_e after Coulomb energy correction, including quantization effects.
- The energy trap depth is derived by matching the experimental and numerical cross points of representations of capture and emission times against gate

bias. The free energy is derived from the activation enthalpy by correcting for the energy barrier for capture E_B and the trap level entropy $T\Delta S$.

- Finally, a comparison of the numerical and experimental thermal dependences fixes the Huang–Rhys factor S_0.

The Coulomb energy has been studied recently for thin gate-oxide transistors (1.7 nm) with a polysilicon gate, including quantization and gate depletion effects [5]. In principle, for a scaled oxide, the Coulomb energy should increase appreciably. Figure 3.3 depicts this situation, resulting in a positive image charge developing on the gate electrode, given by

$$Q_G = q\frac{x_T(C_{inv} + C_D) + C_{ox}t_{ox}}{C_{ox}t_{ox} + t_{ox}(C_{inv} + C_D)} \tag{3.4}$$

C_{inv} and C_D can be computed using a self-consistent Schrödinger–Poisson solver. The oxide capacitance C_{ox} has a component near the gate C_G and near the channel C_c.

For a thick oxide and $x_T = 0$ the results shown in figure 3.3 reduce to the model of Schulz (figure 2.15). As can be seen from figure 3.4, the calculated image charge remains constant until a 2DEG layer appears, whereafter it decreases. The Coulomb energy for small V_{DS} is then given by $Q_G V_{GS}$ [5].

A detailed example of trap parameter extraction has been reported by Amarasinghe *et al* [6], demonstrating that the trap energy remains constant as a function of T with respect to the CB minimum of SiO_2 (figure 3.5(*a*)), while the binding energy of an electron trap decreases with temperature due to the $-T\Delta S$ term. It has also been observed that the capture cross-section prefactor σ_0 increases with gate voltage (figure 3.5(*b*)) [6], which could be a consequence of the fact that the

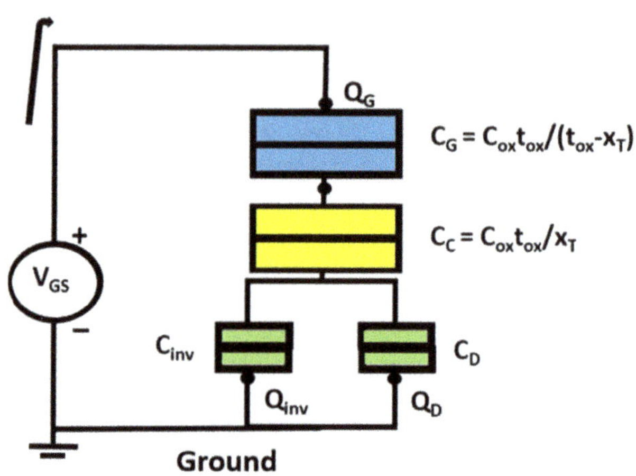

Figure 3.3. Capacitive coupling equivalent circuit, accounting for the effect of trap depth and the charge sharing between the gate, inversion layer, and silicon depletion region. (Reproduced with permission from Lu and Chen [5]. Copyright 2005 the American Physical Society.)

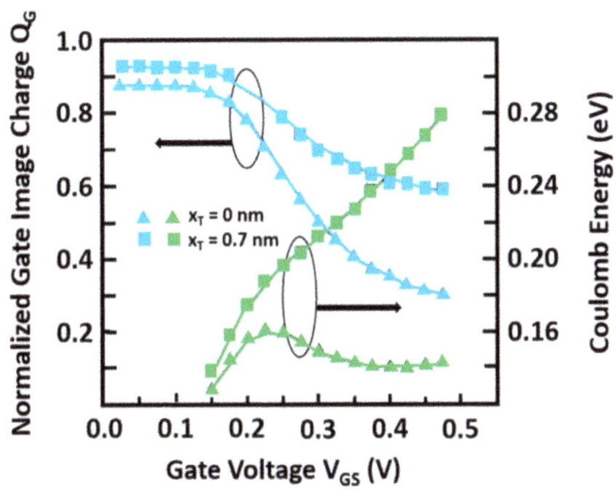

Figure 3.4. Calculated image charge and Coulomb energy versus gate voltage for two trap depths in the oxide. (Reproduced with permission from Lu and Chen [5]. Copyright 2005 the American Physical Society.)

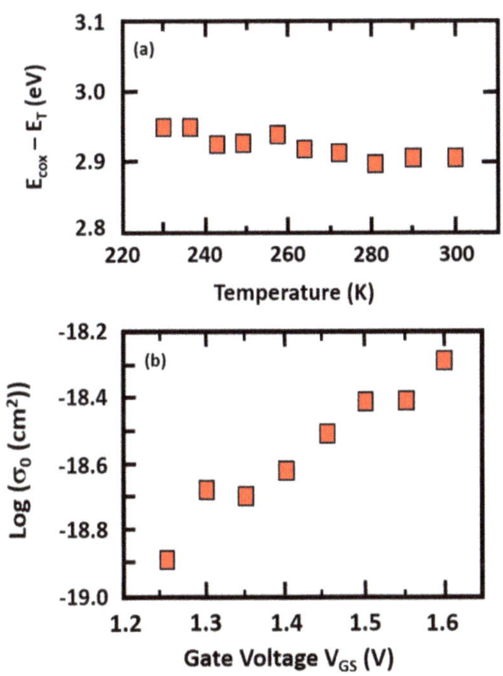

Figure 3.5. (*a*) Temperature dependence of the trap energy with respect to the SiO_2 CB energy ($E_{Cox} - E_T$). (*b*) Gate voltage dependence of capture cross-section prefactor, calculated using the temperature dependence of the mean capture time at each gate bias. (After Amarasinghe *et al* [6].)

inversion layer is squeezed closer to the interface for higher V_{GS}, which increases the tunneling probability. This analysis has been further refined by combining it with a detailed analytical model of the current–voltage characteristics of low doped drain type MOSFETs [7].

A final comment regarding the meaning of the Huang–Rhys factor is that, according to the multi-phonon emission theory, the energy barrier for capture is given by [5]:

$$E_B = \frac{(E_0 - E_T - S_0\hbar\omega)}{4S\hbar\omega} \tag{3.5}$$

in the high-temperature limit, for $kT > \hbar\omega$. $S_0\hbar\omega$ is the lattice relaxation energy depicted in figure 3.6 and is generally larger for a trap deeper in the oxide [5]. At the same time, the prefactor value σ_0 is smaller for a deeper trap, as the wave -function penetration decreases with increasing distance from the interface.

Recently, non-SRH capture has also been found in the inversion mode [8] and accumulation mode [9], i.e. p^+–p–p^+ or n^+–n–n^+, liquid-gated silicon NWFETs, developed as sensor devices. A power law dependence of τ_c on I_D^{-n} with n ~ 2.8 was reported and modeled based on accumulation layer quantization and tunneling

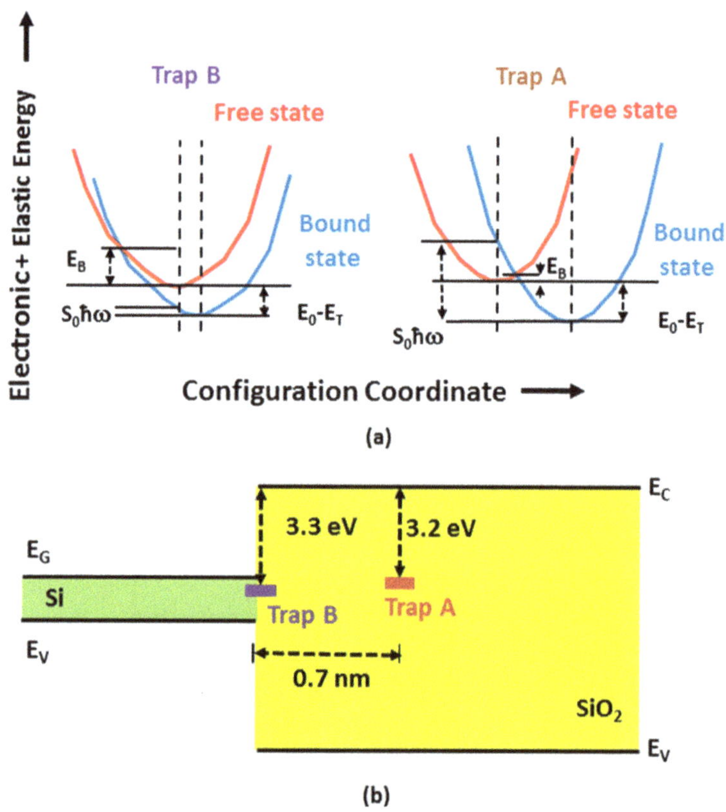

Figure 3.6. Schematic CC diagrams used for a phenomenological description of the capture and emission kinetics encountered in traps A and B. The corresponding energy band diagrams in flat-band conditions are also given, schematically showing the trap depth and its energetic location in the oxide. (Reproduced with permission from Lu and Chen [5]. Copyright 2005 the American Physical Society.)

[8, 9]. It was also demonstrated that inversion (or accumulation) layer quantization affects the depth position of the oxide trap extracted from equations (2.12)–(2.13). This results in a non-negligible correction (order a few tenths of a nanometer), given by [8]

$$x_{\mathrm{T}}^{\mathrm{CL}} - x_{\mathrm{T}}^{\mathrm{QM}} = \beta t_{\mathrm{ox}} \frac{kT}{q} \frac{\partial x_{\mathrm{T}}^{\mathrm{QM}}}{\partial V_{\mathrm{GS}}} \tag{3.6}$$

with $x_{\mathrm{T}}^{\mathrm{CL}}$ and $x_{\mathrm{T}}^{\mathrm{QM}}$ calculated according to equation (2.13) from the classical (CL) and the quantum mechanical (QM) time constant ratios, respectively. The latter is given by [8]

$$\left(\frac{\tau_{\mathrm{c}}}{\tau_{\mathrm{e}}}\right)_{\mathrm{QM}} = \left(\frac{\tau_{\mathrm{c}}}{\tau_{\mathrm{e}}}\right)_{\mathrm{CL}} \frac{1}{T_0} \exp\left(\beta x_{\mathrm{T}}^{\mathrm{QM}}\right) \tag{3.7}$$

with T_0 and β semi-empirical constants [8]: $T_0 = 1$ and $\beta = 0.1\alpha_{\mathrm{t}} = 0.1/\lambda$; α_{t} is the tunneling parameter for the electron (hole) wave-function in the gate dielectric.

A model for the capture time constant has been proposed for RTS occurring *in the GIDL current* [10]. Based on this, an improved theory for the hole capture cross section, considering the multi-phonon emission model, has been derived, enabling the study of the bias dependence of σ_{h}.

It has recently been proposed that studying the RTS parameters as a function of external stress allows deriving the CC diagram; among others, the atomic displacement relaxation energy and the strength of the electron–lattice coupling can be determined [11].

The inverse operation, namely a detailed modeling of the noise spectrum as a function of the trap position has been proposed by Longoni *et al* [12]. It involves numerical device simulation and modeling of the RTS amplitude.

3.2 Extraction trap position from RTS time constants

Several recent papers have developed a detailed strategy to derive the trap depth and lateral position from the RTS time constants [13, 22]. A procedure has also been proposed based on gate-current RTS in a high-κ stack with an IL [23]. Some examples will be discussed here in more detail.

The first case is based on a statistical characterization of I_{D} RTS in 40 nm generation CMOS transistors [19]. Considering the $\tau_{\mathrm{c}}/\tau_{\mathrm{e}}$ ratio of a large number of signals, it was concluded that *two types of traps are present in the gate oxide*, namely, type I communicating with the channel and having a negative slope of $\tau_{\mathrm{c}}/\tau_{\mathrm{e}}$ with V_{GS} (i.e. electron trapping probability increases with V_{GS}) and type II exchanging electrons with the gate electrode and having the opposite slope [19]. This is contrary to the results on older technologies with a thick gate oxide, where at room temperature type-I RTSs have predominantly been observed (see chapter 2). The two types are illustrated schematically in figure 3.7, where the relevant trap parameters are defined [19].

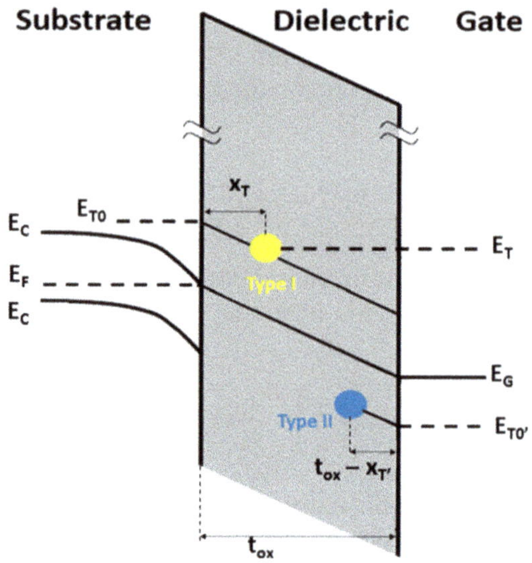

Figure 3.7. Band diagram of a metal–insulator–semiconductor structure with a type-I trap (x_T,E_{T0}) or a type-II trap (x_T', E_{T0}') in the gate dielectric. t_{ox} is assumed to effectively include inversion-layer and poly-depletion thickness. (After Nagumo *et al* [19].)

The depth of the type-I defects can be determined using the standard analysis for strong inversion of equation (2.13). For the type-II traps on the other hand, one can derive the following [19],

$$\frac{t_{ox} - x_T}{t_{ox}} = \frac{kT}{q}\frac{\partial(\ln\tau_c/\tau_e)}{\partial V_{GS}} \tag{3.8a}$$

$$E_{T0} - E_G = -q(V_{G0} - V_0)\left[1 - \frac{x_T}{t_{ox}}\right] \tag{3.8b}$$

It should be remarked that the RTSs have been measured above V_T, so that one can probably neglect the Coulomb blockade effects discussed in section 2.1.3, which could also cause a type -I time constant ratio behavior. Based on the above described approach, most of the RTSs appear to be grouped in the two (energy, depth) windows represented in figure 3.8, whereby no correlation has been observed between the trap time constant and its depth, nor between the amplitude and the depth [19].

Most of the analysis so far has been derived for a single-gate dielectric with uniform parameters (SiO_2 or SiON). In the case of devices with a high-κ gate, there is generally a thin SiO_2 IL in order to reduce the density of interface states with the silicon substrate. Recently, cap layers based on other metal oxides have also been introduced to perform so-called work function (WF) engineering. This implies that the gate stack may have different properties (κ-value, trap density) with depth. This means that the models for the extraction of the trap depth based on the I_D or I_G RTS time constant ratio have to be adapted, taking into account the fact that a defect

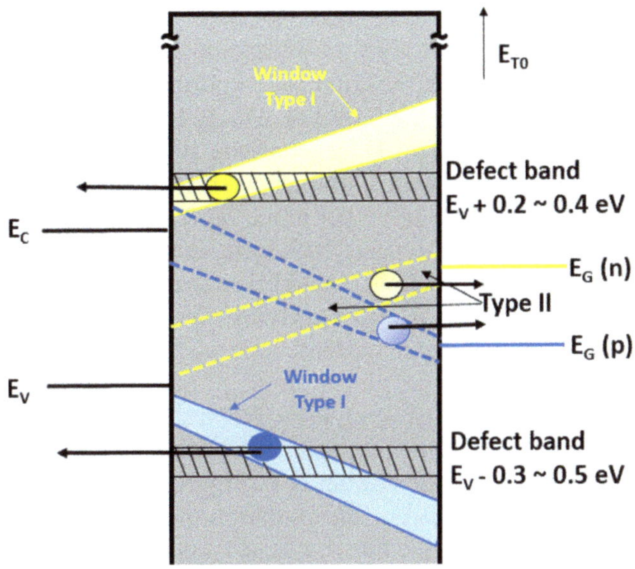

Figure. 3.8. Representation of the obtained trap distribution in the gate dielectric (SiO_2) and the window in the $x_T - E_{T0}$ plane. (After Nagumo *et al* [19].)

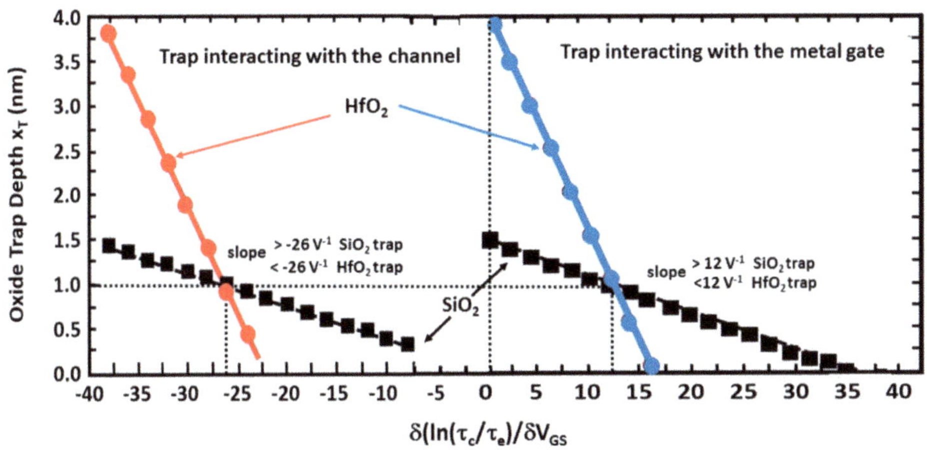

Figure 3.9. Oxide-trap depth corresponding to the gate voltage derivative of the RTS time constant ratio for a 1 nm SiO_2 and a 3 nm HfO_2 bilayer stack for a trap interacting with either the channel (left) or the TiN metal gate (right). (After Cho *et al* [26].)

may be in the SiO_2 IL or in the high-κ layer. Appropriate expressions have recently been worked out and can be found in the literature [23–25]. Based on these, a simple way to discriminate a trap in the IL from one in the high-κ layer has been proposed and is represented in figure 3.9 for a trap interacting with, respectively, the channel [24] and the metal gate [26].

In the case of a trap interacting with the channel (the left part of figure 3.9) it can be seen that the traps in the high-κ layer have a more negative slope ($<-26\ V^{-1}$) than a defect in the IL. Moreover, this analysis can be extended to the case of I_G RTS [24–27], where traps in the high-κ layer may communicate either with the channel or with the gate electrode. Again, in the latter case a positive τ_c/τ_e slope is considered as a fingerprint of trapping from the gate [24–27]. It has been concluded that it is difficult to observe a drain current RTS stemming from a trap in the high-κ layer because there is a competition with simultaneous thermionic emission to the oxide CB of a trapped charge [24]. On the other hand, I_G RTN allows one to detect traps in the high-κ layer which communicate with both electrodes [24], emphasizing the power of this complementary method. In addition, a method to analyze RTS traps in the *gate edge overlap region* of high-κ nMOSFETs has been proposed, whereby the signal only appears in the gate current, with no signal in the drain current in inversion mode [28].

Another recently opened avenue of RTS research is the study of MOSFETs biased in accumulation which, in principle, carry no channel current but mainly GIDL, gate current and substrate current [28]. Figure 3.10 shows two examples of RTS in a high-κ MOSFET biased in accumulation mode [30]. In the first case, the RTS is only in I_D (GIDL) and I_B, while in case (*b*) it is in I_G and I_D. This suggests in the latter example that the fluctuations are in the gate edge direct-tunneling (EDT) current that flows between the gate and the drain (or source) overlapped by the gate. This EDT component becomes more pronounced for thinner gate oxides. It has been shown that an EDT current RTS can be identified by the fact that the emission time

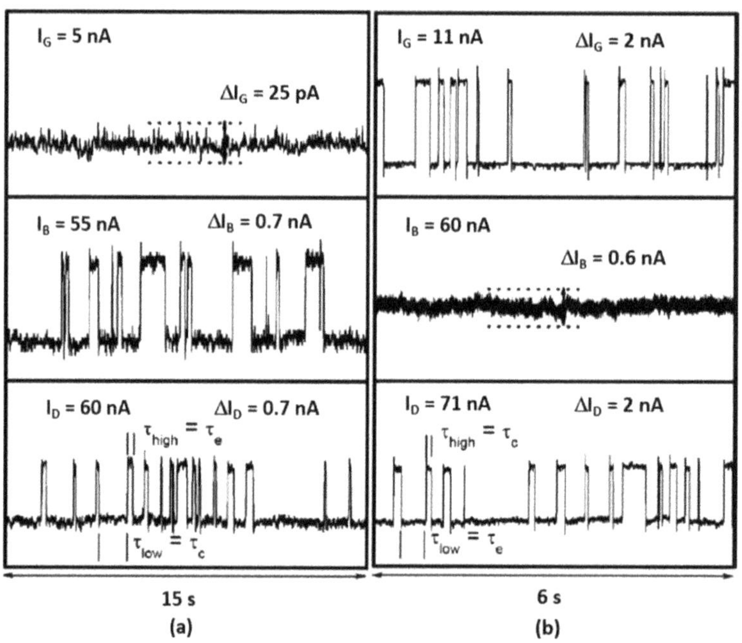

Figure 3.10. Measured I_D, I_B and I_G values under GIDL bias conditions of $V_{GS} = -0.3$ V and $V_{DS} = +1.5$ V in nFETs with L values of (*a*) 0.3 and (*b*) 0.08 μm. The currents are measured one after the other. The FETs in (*a*) and (*b*) show the RTS in the GIDL current only and in the EDT current only, respectively. (Reproduced with permission from Lee *et al* [30]. Copyright 2010 IEEE.)

constant increases with gate voltage (nMOS), while the capture time decreases [30]. For a GIDL current RTS, τ_c is independent of V_{GS}, while τ_e increases with gate voltage [30]. By using a procedure to separate the EDT current in the source and drain, one can also separate the different components of the RTS amplitude, providing further insight into the matter [31].

More recently, the same authors have demonstrated that there is no correlation between ΔI_B (due to the GIDL current) and ΔI_G (due to the EDT current) [32]. This means that traps in the overlapped region only affect one of the two current components, not both [32]. The different cases are illustrated by figure 3.11. In addition, according to both experimental and simulation results, there is no correlation between the trap depth x_T and the current amplitude for GIDL.

RTS in the gate current under accumulation bias has also been studied for devices with a high-κ dielectric stack [26]. One can again discriminate gate and channel traps based on the sign of the time constant ratio with V_{GS}, as shown in figure 3.9. The right part of this figure is valid for a trap interacting with the metal gate. A positive slope indicates a trap which interacts with the gate. Based on the value of the slope, it is possible to identify traps in the SiO_2 IL (>12 V^{-1}) or in the HfO_2 (<12 V^{-1}) [26]. The slope has not only a different sign, but also a different magnitude. At the same time, it has been shown that gate edge damage can introduce RTS in the gate EDT current [33]. It was derived from the time constants that the plasma-induced oxide traps were found to be distributed over a wide area of the high-κ/SiO_2 sidewall at the gate edge region.

An alternative model for RTS in the GIDL current considers the impact of a charged or empty oxide trap on the carrier generation of a nearby fast interface trap

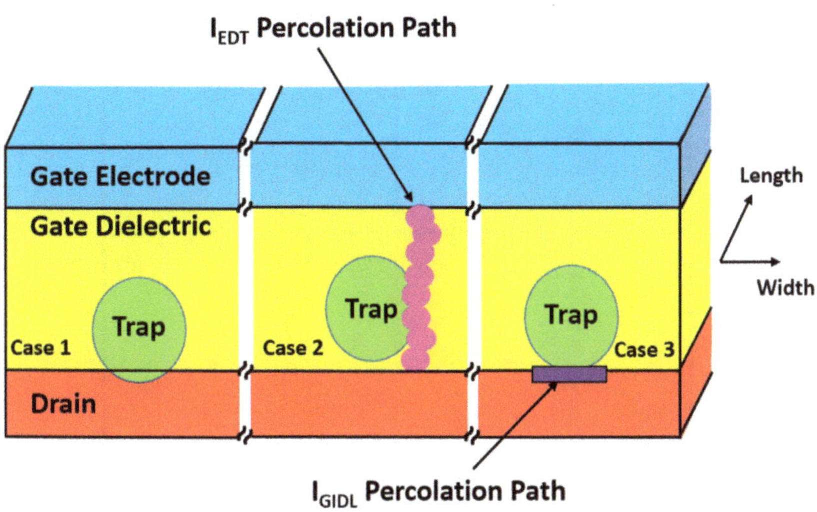

Figure 3.11. A schematic cross-sectional view showing three different traps located in the gate dielectric on the drain overlapped by the gate along the width direction. Case 1 shows that a trap is located in the gate dielectric, and no defect is located within several Debye lengths from the trap. Case 2 shows that a trap is located at a position near an I_{EDT} PP through the dielectric. Case 3 shows that a trap is located inside the gate dielectric on an I_{GIDL} PP (a localized region at the surface of the drain). (After Lee and Lee [32].)

at a distance r [34–36]. In that case, the trap depth can be expressed in terms of the distance with respect to the fast trap and considering the electric field at the interface trap [36]. A different method is to consider contour maps of the relative RTS amplitude $\Delta I_{GIDL}/I_{GIDL}$ at fixed drain-to-gate voltage V_{DG} to derive x_T in the gate-to-drain overlapped region [34].

Quite recently, the impact of the electrostatics and the doping-induced variability on the RTS time constants in nanoscale MOSFETs has been discussed [37–39]. It has been shown that both factors contribute significantly to their statistical dispersion. The dispersion in the capture time constant is a direct consequence of the non-uniform inversion channel, i.e. of the local electron concentration below the oxide trap. This is illustrated by figure 3.12, representing τ_c as a function of trap position along the channel and showing slightly higher values near the drain at small V_{DS} (100 mV). In addition, at fixed L the scatter in τ_c occurs along the width direction due to atomistic doping variations [37]. Atomistic doping represents as strong a variability source of τ_c as the 3D electrostatics along the channel length. The τ_c variability decreases in strong versus weak inversion.

According to figure 3.13, there exists a nearly perfect inverse relation between τ_c and $1/n$, i.e. with n at the trap position [37]. On the other hand, the variability in the tunneling transparency is negligible for all x_T depth positions. At the same time, the spread in emission time constant τ_e is much smaller in figure 3.13 [37]. This spread is positively correlated with the spread in τ_c and becomes smaller in strong inversion

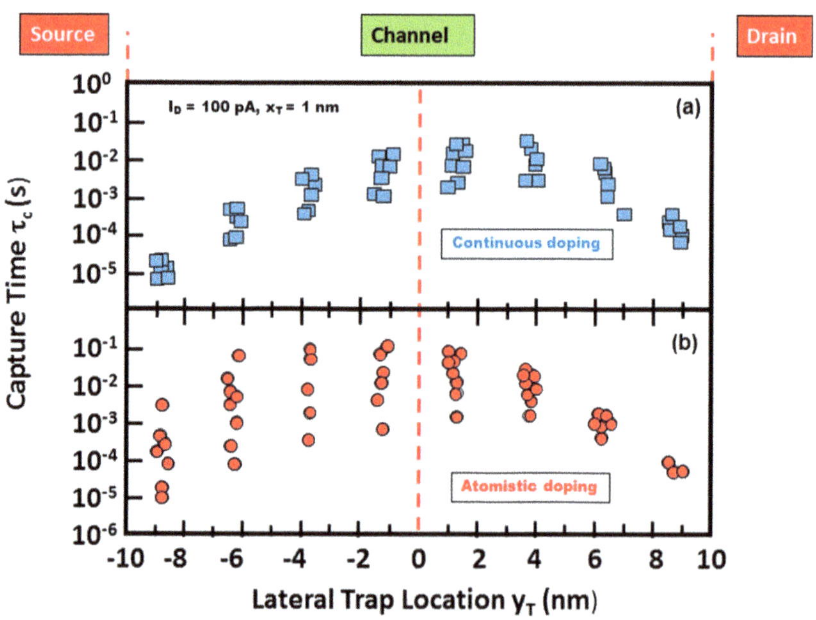

Figure 3.12. Calculated τ_c as a function of trap position along L for a trap depth x_T = 1 nm and I_D = 100 pA. Results for (*a*) continuous and (*b*) one case of atomistic substrate doping are shown. The channel center, source and drain edges are at 0, −10 and +10 nm, respectively. Each point in the figure corresponds to one of the 64 trap positions uniformly selected along the channel area. (After Castellani *et al* [37].)

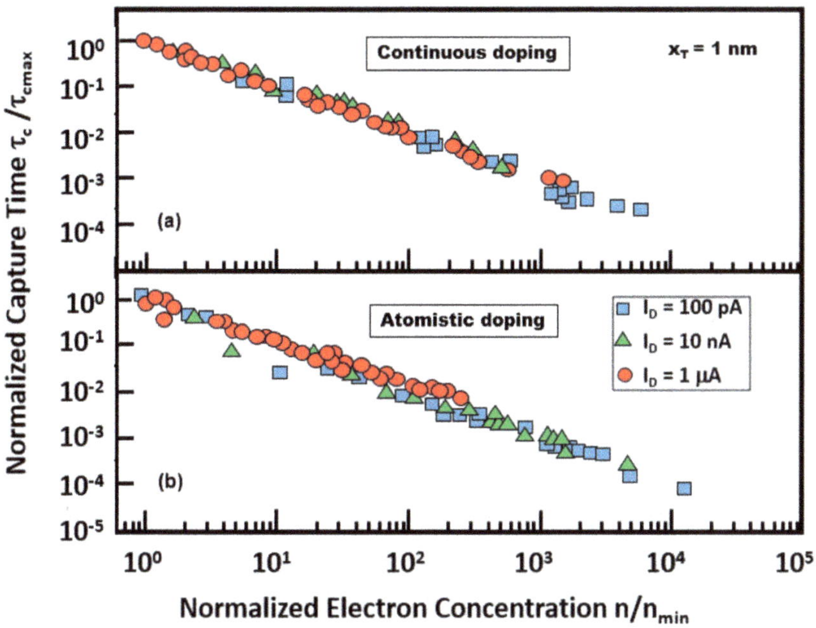

Figure 3.13. Calculated τ_c for traps at $x_T = 1$ nm as a function of n, for $I_D = 100$ pA, 10 nA and 1 μA. Results for (*a*) continuous doping and (*b*) the atomistic cell are shown. (After Castellani *et al* [37].)

due to electrostatic screening. Moreover, quantization corrections have a negligible impact.

The statistical spread in the time constants (in particular to τ_c), spanning for example several orders in figure 3.14, seriously questions the extraction of the trap depth x_T for $L = 20$ nm transistors, based on the time constant ratio [38]. Atomistic Monte Carlo (MC) simulations demonstrate that the maximum error in x_T extraction in that case occurs when the trap is right on top of a dopant atom in the channel. The experimental assessment of x_T using, e.g. equation (3.7) has limited validity (and accuracy) in the presence of 3D electrostatics (short-channel effect) and atomistic doping. This is illustrated by figure 3.15, showing the large spread in extracted trap depth [38].

At the same time, it is shown that there exists little correlation between the amplitude, expressed as ΔV_T and the trap depth and time constants. This is due to the large statistical spread coming from percolative channel conduction, particularly in weak inversion [37]. This is demonstrated by figure 3.16 [38]. In principle, one would expect a negative correlation between ΔV_T and τ_c, which is not found in practice [40]. This implies that 3D electrostatics and atomistic doping affect the RTN amplitude and time constants in a different way, resulting in no correlation in the spread of ΔV_T and τ_c [40]. In addition, no correlation between ΔV_T and n is demonstrated, while a weak correlation between ΔV_T and x_T is found in figure 3.16 [38].

Also, the impact of the metal gate granularity (MGG) and the line edge roughness (LER) on the accuracy of x_T has been evaluated [39], showing a significant impact of

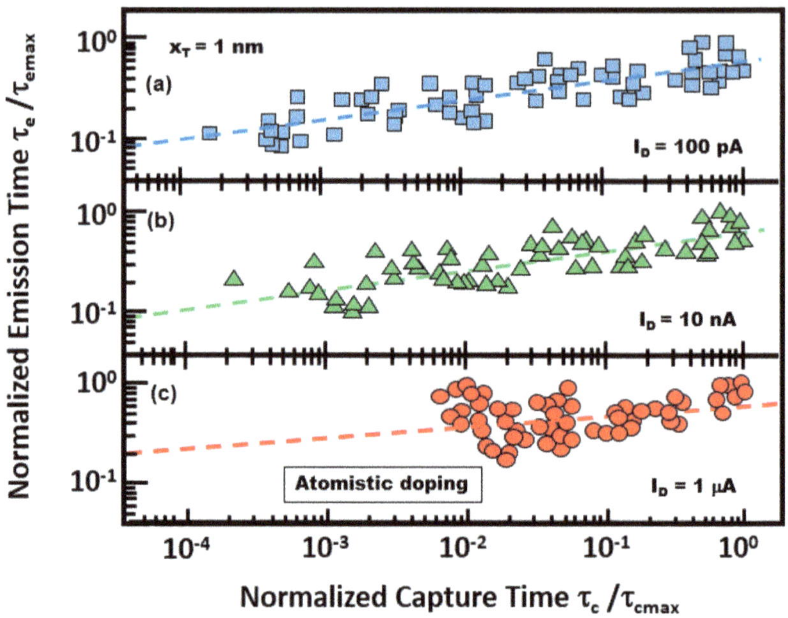

Figure 3.14. Calculated τ_e as a function of τ_c (both normalized to their maximum value) for the atomistic cell and x_T = 1 nm, in the case of *(a)* I_D = 100 pA, *(b)* 10 nA and *(c)* 1 μA. (After Castellani *et al* [37].)

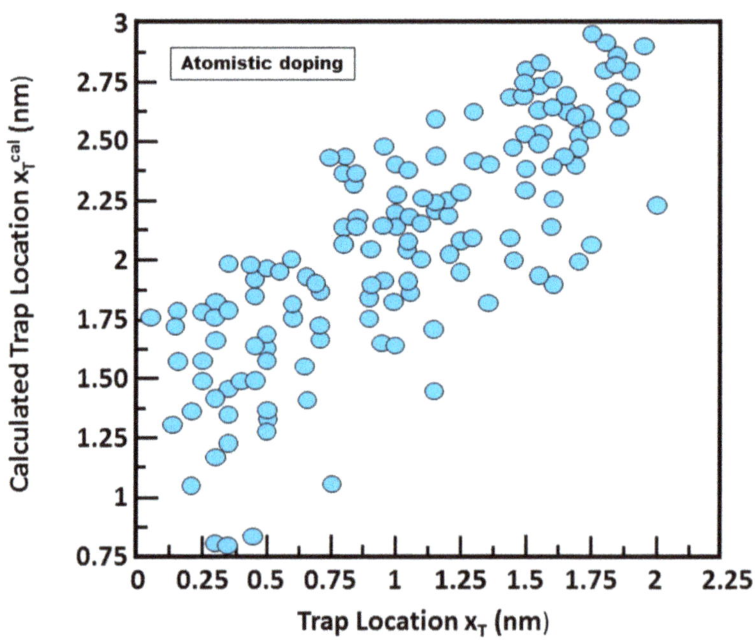

Figure 3.15. x_T^{calc} as a function of x_T for many cells having different atomistic configurations of substrate doping and a single RTN trap randomly placed over their channel area, and in the x_T range between 0 and 2 nm. (After Monzio Compagnoni *et al* [38].)

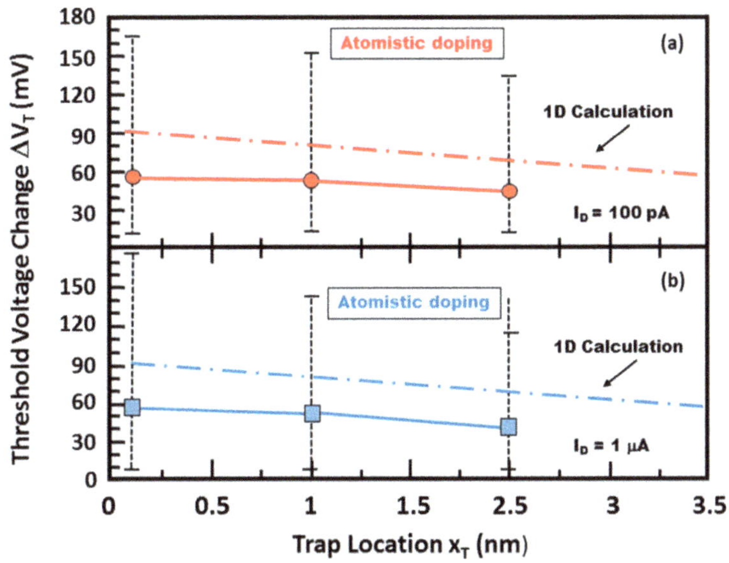

Figure 3.16. Average ΔV_T for the 64 RTN traps uniformly selected over the channel of a single atomistic cell, as a function of x_T and in the case I_D = 100 pA (*a*) and 1 μA (*b*). The error bar highlights the maximum and minimum ΔV_T values for the traps. Results of 1D calculations are also shown. (After Monzio Compagnoni *et al* [38].)

MGG, while LER has a negligible contribution. It has been demonstrated that high errors in x_T result for high values of the metal WF, due to the lower electron concentration at the defect site. In general, the calculated trap depth is larger than the real position (figure 3.15). On the other hand, for thin gate oxides, the trap may also interact with the gate (emission to the gate), whereby the probability for this to occur increases with gate voltage [39]. As a result, τ_{ctot}/τ_{etot} exhibits a much weaker dependence on V_{GS}, resulting in a calculated trap position too close to the Si/SiO$_2$ interface.

The above considerations have inspired the research group at TU Wien to establish a more robust extraction technique for the lateral oxide-trap position from the time constant ratio, including RDF in the channel [41–43]. It is based on the dependence between the trap-induced surface-potential shift in the perturbed region $\delta\Psi_s^T$ and V_{DS}. Quantum corrections have been implemented to account for the electron quantum correction potential [42]. This is done in order to minimize the dependence of the local surface potential near the trap on the grid meshing. The quantum corrections only weakly depend on V_{DS} and have no marked impact on the differential $\delta\Psi_s^T(V_{DS})$ characteristic. In first approximation, the trap position can be found from the polynomial function $p(y_T)$, which is defined by

$$\delta\Psi_s^T(y_T, V_{DS}) - \delta\Psi_s^T(y_T, V_{DS} = 0\text{ V}) = p(y_T)V_{DS} \tag{3.9}$$

In equation (3.9), $p(y_T)$ is hardly affected by the random dopant fluctuations, in particular for longer channels [42], while it strongly increases towards the drain. It should be noted that $\delta\Psi_s^T(y_T, V_{DS} = 0\text{ V})$ may vary for different devices, corresponding to different random dopant placements. This implies that only the V_{DS}-induced shift can be employed for trap determination in a reliable way [42].

In practice, the V_{DS}-induced threshold voltage shift ΔV_T is used for the derivation of the lateral trap position [41]. The RDF leads to some uncertainty. In the first instance, the $\Delta V_T(V_{DS})$ dependence is parametrized by a cubic function of V_{DS}, whereby the corresponding coefficients are unique for each trap position along the channel, while the effect of the RDFs is much weaker [41]. In the modeling, the trap potential peak is represented by a Voigt-like peak function given by [41]

$$\delta \Psi_s^{T0}(y) = \pm \frac{V_0}{1 + \left(\frac{y - y_T}{y_0}\right)^2} \tag{3.10}$$

with the + for a pMOSFET and the – for an nMOSFET. The height of the potential spike V_0 is independent of y_T and taken between 0.4 V and 1 V, with $x_0 = 1$ nm.

It has been found that the traps in the middle of the channel are responsible for a much larger doping shift than those close to the S/D electrodes. In addition, it is shown that the trap-induced ΔV_T increases with V_{DS} near the source, while it decreases near the drain [36]. In the middle of the channel, a weak dependence is found, indicating that the accuracy of the method will be smallest for traps located there [41]. The impact of RDF on the determination of the lateral trap position increases with increasing substrate doping density and falls in the range 2 to 5 nm, for an average substrate doping density or an off average doping transistor [41].

3.3 RTS amplitude modeling

In the past, three main approaches have been followed to model the RTS amplitude [44], assuming, respectively:

- Uniform charge-carrier and mobility fluctuation.
- A local perturbation embedded in a uniform channel.
- A local perturbation in a non-uniform channel.

The first class of models has been addressed in section 2.2. As mentioned before, while this provides some qualitative insight, it is difficult to understand RTS amplitudes larger than 1% [45] and moreover, to explain the wide range of amplitudes in the same device. The second approach provides a better insight into the parameters affecting the RTS amplitude. In the simplest case, such as for the channel current RTS model of figure 3.17, it is considered that the local conductivity around a cored-out trap area σ_t is different from the average channel conductivity σ when the trap is charged [46–48]. The impact of the charging of the trap can immediately be evaluated if the conductivity σ_t in a square area with side L_t (trapping length) around it is zero. Solving the resistance network in figure 3.17(*b*), immediately leads to [47]

$$\frac{\Delta I_D}{I_D} = \frac{L_t^2(\sigma - \sigma_t)}{WL\sigma - L_t(L - L_t)(\sigma - \sigma_t)} \approx \frac{L_t^2}{WL} \tag{3.11}$$

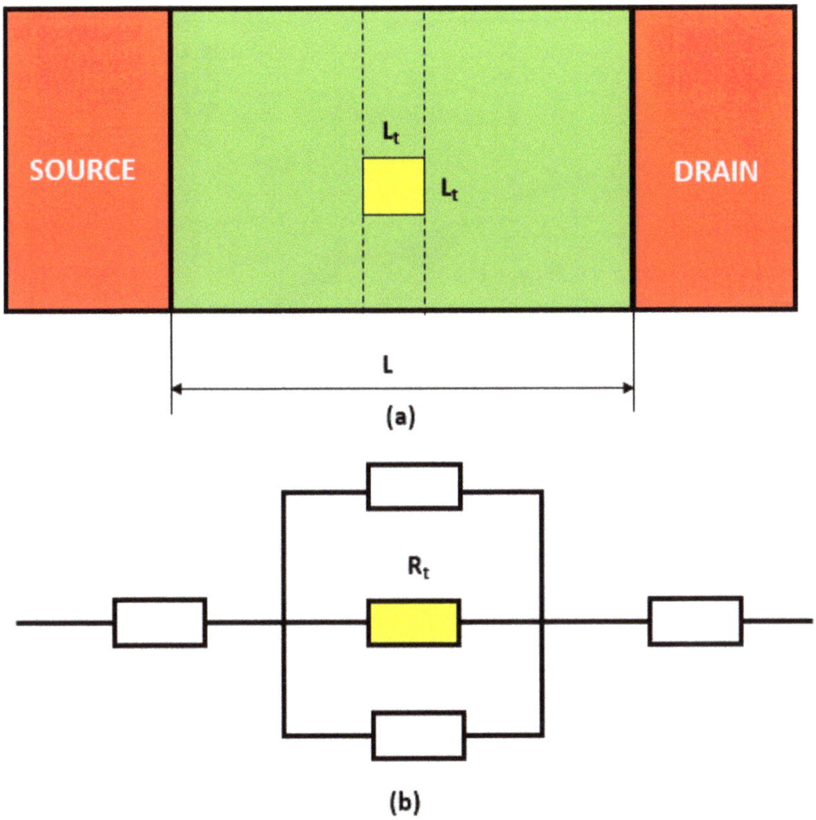

Figure 3.17. An interface trap in a MOSFET with area $W \times L$ (*a*) and the equivalent parallel resistance network (*b*), used in the calculation of the channel resistance modulation ΔR.

Several approximations for L_t have been derived in the literature [47, 49]. For example, it has been proposed to model L_t by [49]

$$\frac{1}{L_t} = \frac{\gamma_1}{L_S} + \frac{\gamma_2}{L_C} \tag{3.12}$$

L_S is twice the critical distance from the trapped electron where the Coulomb potential energy is equal to the carrier average kinetic energy kT (no carrier heating) in a 2DEG and can be written as $q^2/2\pi\varepsilon_{\text{eff}}kT$ where ε_{eff} is the effective permittivity, which is in the first instance given by $(\varepsilon_{\text{Si}} + \varepsilon_{\text{ox}})\varepsilon_0/2$ [50]. L_C is the screening length by the 2DEG and, in the quantum limit, it can be written as $2^{3/2}\varepsilon_0\varepsilon_{\text{Si}}kT/q^2n_s$. In this approximation, the screening by the (metal) gate is neglected, which becomes important for thin gate oxide devices [50]. In equation (3.12), γ_1 and γ_2 ($\leqslant 1$) are coefficients accounting for the percolation areas in the subthreshold and inversion regimes of operation, respectively. For $\gamma_1 = \gamma_2 = 1$, the uniform channel approach of [47] is retrieved. The resulting calculated trap length is represented in figure 3.18, showing good agreement with the measured RTS amplitudes as a function of V_{GS} [49].

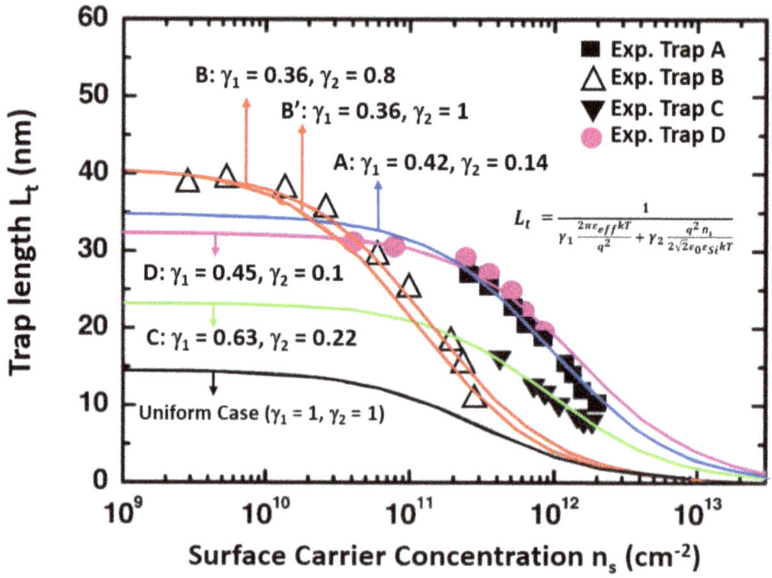

Figure 3.18. Comparison of experimental Debye length (symbols) versus 2DEG carrier density with those calculated (lines) from equation (3.12). Also shown is the calculated line for the uniform case. (Reproduced with permission from Chen *et al* [49]. Copyright 2008 AIP Publishing).

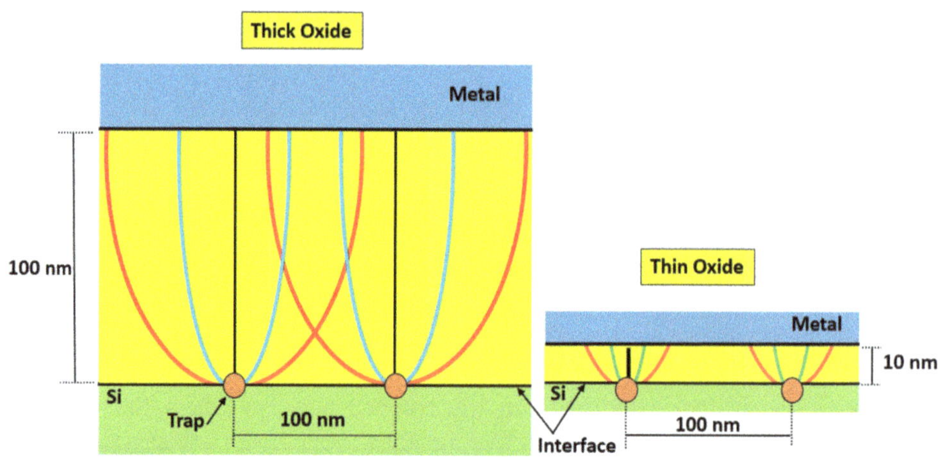

Figure 3.19. Schematic of screening of interface point charges due to the proximity of the gate electrode. In the case of a thick oxide with $t_{ox} = 100$ nm, the regions of impact of charged traps, visualized as the field lines due to the presence of localized point charges in the interface, overlap for a trap distance of 100 nm. In the case of a thin oxide ($t_{ox} = 10$ nm), traps with the same mean distance are effectively isolated. (After Mueller and Schulz [44].)

A further refinement of this local conductance model in a non-uniform channel considers also the impact of the gate screening on the spatial extent of the potential perturbation due to a localized interface charge in the channel [44, 50], schematically represented by figure 3.19. The proximity of the gate electrode causes considerable screening of the potential in the plane of the interface.

For an interface trap density of 10^{10} cm^{-2}, the average trap separation is 100 nm. In the case of a thin oxide, the regions of impact of neighboring traps are effectively separated in figure 3.19 (right), the potential overlap is screened by the gate electrode. The maximum of the radial charge is located at approximately half t_{ox}, both below and above the threshold. In this case, the (Debye) screening only marginally depends on the free carrier density in the channel and is thus electrostatic in nature. RTSs with such a density can be considered to *act independently* (or be uncorrelated) due to the gate screening. According to figure 3.20, the area density of the channel charge induced by the trapped charge at the interface drops to 10% of its maximum value one oxide thickness away from the trap (at the origin). It is clear that the region significantly perturbed by one fixed charge is less than one oxide thickness. It was also derived that the mean free carrier path derived from mobility is of the same order as the trap cross section $2\pi r$ for scattering [51].

In the worst case scenario in which the trap blocks all current flow inside a disk area of radius r, i.e. it cores out a disc of zero conductivity when charged, the corresponding conductivity change is given by [44]

$$\frac{\Delta G}{G} = \frac{4}{\pi}\frac{\text{area (disk)}}{\text{area (channel)}} \tag{3.13}$$

which is, up to the geometrical factor $4/\pi$, equivalent to equation (3.11). Assuming a radius in the range of 4 to 10 nm ($\sim 0.8t_{ox}$) yields a maximum RTS amplitude of 0.5%. The cut-out disc picture is appropriate below and at threshold, as long as the mobile charge density in the uniform channel is smaller than 1/(disc area). In strong

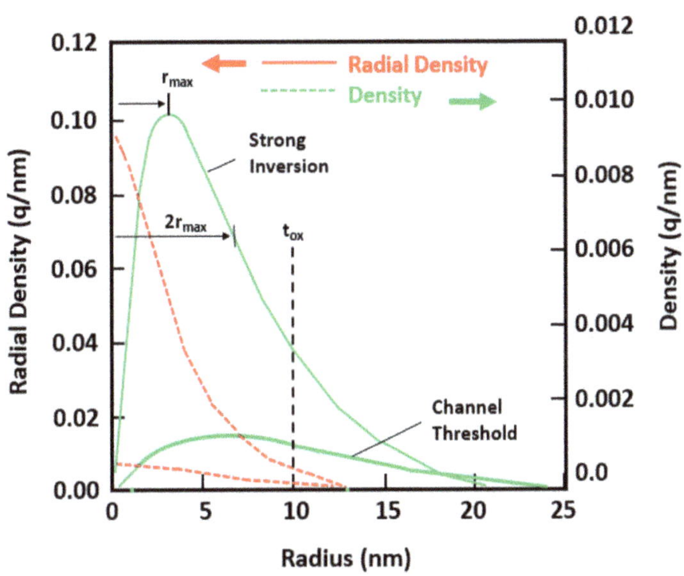

Figure 3.20. Channel charge induced near a fixed interface point charge. The radius r is the distance in the plane of the channel from the location of the fixed charge. Solid curves denote the radial charge density $2\pi rq\Delta n_s(r)$; dotted curves the area charge density $q\Delta n_s(r)$ at threshold and in inversion as indicated. (After Mueller and Schulz [44].)

inversion, the cut-out disc still contains charge, so that the conductivity is not zero. Thus, above threshold, $\Delta G/G$ will decrease due to $G_{disc} > 0$ and due to the reduction of the radius by screening, explaining the $1/G$ roll-off in inversion [44].

The case of an attractive trap can be modeled in the same way, with a disc with increased conductivity when charged, in the limit $G = \infty$ [44]. The corresponding amplitude is represented in figure 3.21. According to this picture, no quantitative differences are expected between the RTS amplitude caused by either attractive or repulsive traps [44]. Based on this approach, and assuming a non-uniform channel, a method has been developed to use the RTS amplitude as an atomic-size probe for the local current in the channel [51]. This enables one to quantify the channel non-uniformity due to, e.g., surface-potential fluctuations induced by fixed oxide charges, which may be helpful in calibrating numerical modeling.

The third set of models, adding an additional level of numerical sophistication and complexity, considers a non-uniform conductance change in a non-uniform channel, whereby the surface-potential fluctuations are induced by fixed oxide charges [44]. This leads to current percolation (or current filamentation). This approach successfully reproduces RTS amplitudes above 1% [44], explains the weak temperature dependence of the RTS amplitude, shown in figure 3.22 [44], and yields a satisfactory agreement with the observed distribution in amplitudes (figure 3.23) [51]. A percolation model for RTS was recently also proposed in reference [52] and has received support from data on large memory arrays (see chapter 6).

As will be shown more extensively in the following paragraphs, the RTS amplitude is widely distributed over several orders. Attempts have been made to

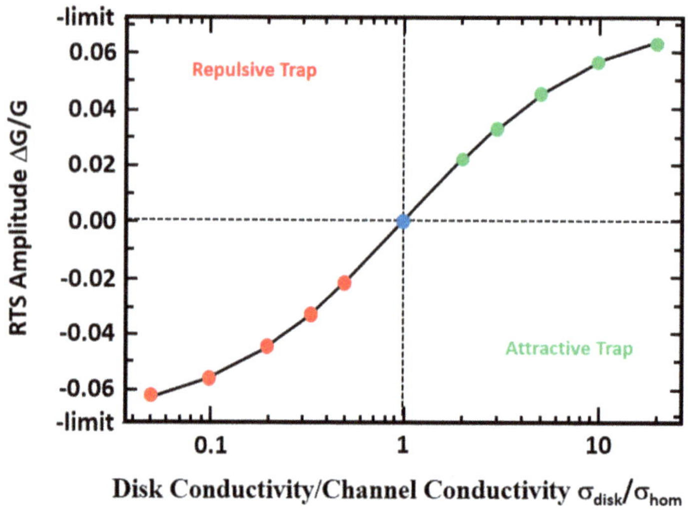

Figure 3.21. Relative conductance modulation $\Delta G/G$ caused by a disc (radius 50 nm) of conductivity σ_{disc} introduced in the center of the homogeneous model channel (conductivity σ_{hom}). The left-hand side, where $\sigma_{disc}/\sigma_{hom} < 1$, describes the model for the repulsive trap; the right-hand side ($\sigma_{disc}/\sigma_{hom} > 1$) the model for the attractive trap. The top and bottom borders of the plot denote the symmetric limits of $\Delta G/G$ for $\sigma_{disc}/\sigma_{hom} \to \infty$ and $\sigma_{disc}/\sigma_{hom} \to 0$, respectively. (Reproduced with permission from Mueller and Schulz [44]. Copyright 1996 AIP Publishing.)

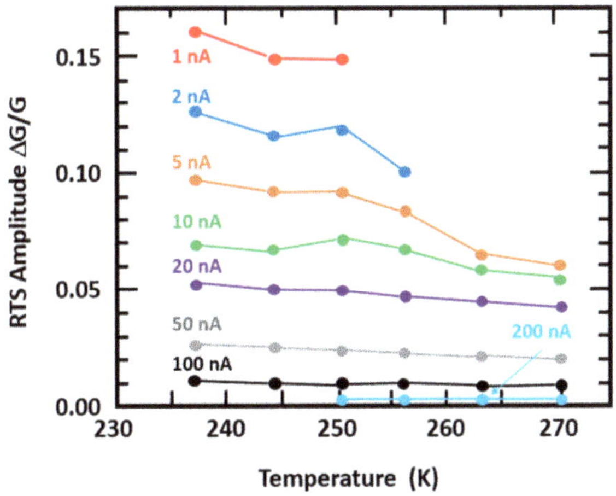

Figure 3.22. Relative RTS amplitude $\Delta G/G$ for one specific interface trap versus sample temperature T. Parameter is the drain current I_D at constant drain voltage $V_{DS} = 10$ mV. (Reproduced with permission from Mueller and Schulz [44]. copyright 1994 AIP Publishing.)

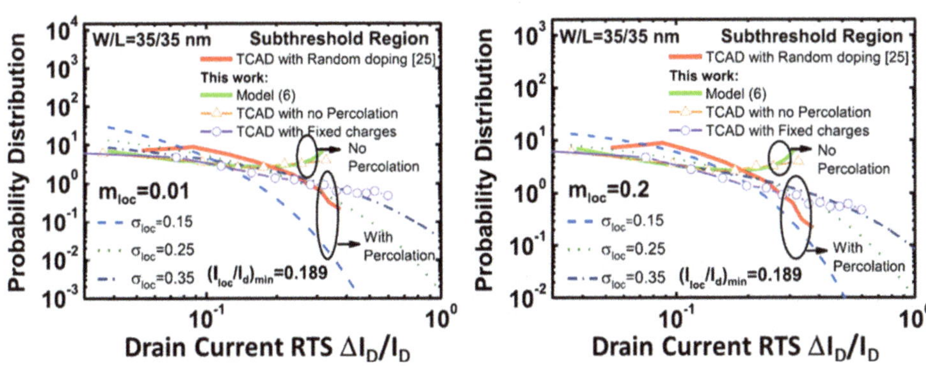

Figure 3.23. Calculated $\Delta I_D/I_D$ headed and tail distribution curves. $L_{t0} = 19.7$ nm, $a = 0.10$ and $b = 4.11$ nm. (*a*) $m_{loc} = 0.01$ and (*b*) $m_{loc} = 0.2$, with σ_{loc} as a parameter. Technology computer-aided design (TCAD) created distributions at $V_{DS} = 0.01$ V are also included and shown for comparison. Here $V_{GS} = 0$ V and $V_{DS} = 0.05$ V. (Reproduced with permission, from Chen *et al* [53]. Copyright 2014 IEEE.)

predict these distributions and in particular its head and tail parts [53, 54]. The head tail is associated in this work with the uncertainty of the trap position without percolation and the high-amplitude tail corresponds to the presence of percolation. A general expression for the RTS amplitude is given by [53]

$$\frac{\Delta I_D}{I_D} = \frac{I_{loc}}{I_D} \tag{3.14}$$

with I_{loc} the local current around the trap and I_D the average channel current. The average m_{loc} and standard deviation σ_{loc} of the variable I_{loc}/I_D reflect the underlying individual PP due to the use of the same manufacturing process [53]. Representing the normalized RTS amplitude by equation (3.11), one can formulate a closed-form description for a headed distribution without percolation. The trap length L_t can be derived from device simulations for a negatively charged trap along the channel direction in the middle of the device width. This is based on the argument that the relative RTS amplitude only weakly changes along W [53]. From the calculations, an analytical model for the trap length can be fitted as a function of the distance d from the mid-channel, given by [48]

$$L_t = \frac{L_{t0}}{\sqrt{1 + a\left(\exp\left(\frac{d}{b}\right) - 1\right)}} \tag{3.15}$$

with $L_{t0} = 19.7$ nm for a mid-channel trap in the process condition studied [53], while $a = 0.1$, $b = 4.11$ nm. Based on equations (3.14) and (3.15) the following probability distribution function for the headed distribution has been established [53]:

$$g\left(\frac{\Delta I_D}{I_D}\right) = 2bf(d)\left(1 + \frac{1 - a}{a}\exp\left(-\frac{d}{b}\right)\right) \Big/ \frac{\Delta I_D}{I_D} \tag{3.16}$$

which yields a satisfactory approximation for the experimental headed distribution, as shown in figure 3.23. It is found that the calculated distribution function of $\Delta I_D/I_D$ piles up near the upper limit (mid-channel trap), forming the head of the distribution.

It has been argued that the lower $\Delta I_D/I_D$ distribution in the presence of percolation is considerably close to that of the headed one. The lower limit of I_{loc}/I_D can be reasonably set by the lower limit of $\Delta I_D/I_D$ with trap position $d = L/2$ [53]. Two types of RTSs can be identified in the case of percolation: strategic ones which contribute to the tail, and non-strategic ones which repopulate the non-percolative head part. Thus, part of the headed distribution near the upper limit is repopulated and separated into two components in the percolation case: a higher $\Delta I_D/I_D$ part and a lower $\Delta I_D/I_D$ part. The required intersection between the non-percolative and the percolative distribution serves as a second criterion to judge the validity of the numerically established distributions [53].

Figure 3.24 represents σ_{loc} versus m_{loc}, showing that these parameters are not unlimited. The extracted m_{loc} and σ_{loc} from a percolation channel are not unique. In other words, the tail distribution of a percolative channel can be reproduced from a few critical m_{loc} and σ_{loc} values [53]. In addition, different percolation origins (fixed charges, random dopants, LER, gate granularity, etc) can result in different combinations of m_{loc} and σ_{loc} [53].

Finally, figure 3.25 represents the extracted L_t as a function of the lateral trap position with respect to mid-channel ($d = 0$) for different V_{GS}. Two clear trends are obvious: the maximum L_t is found for a mid-channel trap and L_t decreases with increasing gate voltage for an nMOSFET. Based on this theoretical approach, a graphical method has been developed to analyze the ΔV_T induced by RTS in a percolative channel [54].

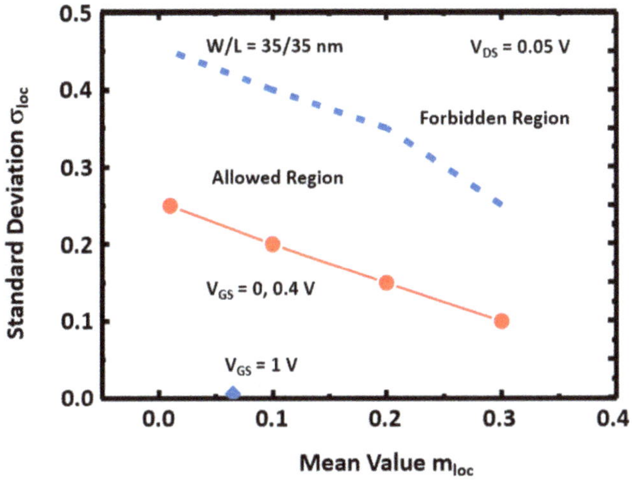

Figure 3.24. Plot of the critical σ_{loc} versus m_{loc} curve. Only for sets of m_{loc} and σ_{loc} lying below the curve can the calculated tail distribution have intersections with the headed distribution. Extracted σ_{loc} and m_{loc} from the tail distributions for three gate voltages appear in the allowed region. (Reproduced with permission from Chen *et al* [53]. Copyright 2014 IEEE.)

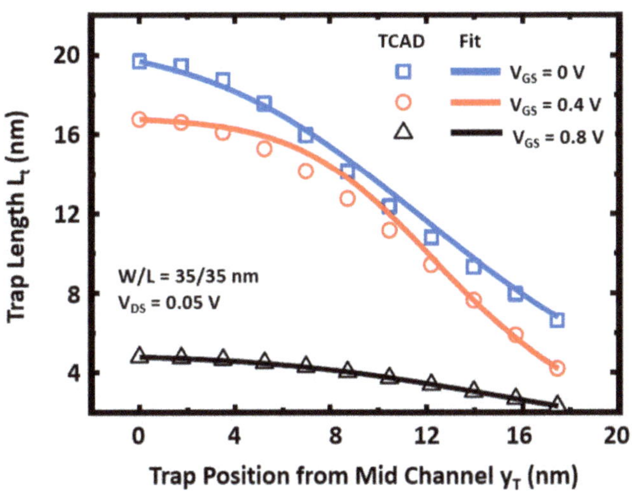

Figure 3.25. Symbols: extracted L_t versus the trap distance ($y_T = 0$ in the middle of the channel between source and drain) at three gate voltages. (Reproduced with permission from Chen *et al* [53]. Copyright 2014 IEEE.)

3.4 Atomistic numerical modeling of the RTS amplitude

It is well-established by now that transistor scaling to the sub-100 nm range (decananometer) brings about issues related to RDFs. The random location of a small number of dopant atoms in a short channel first of all leads to local surface-potential variations, similar to the fixed oxide charge and, in addition, results in a large(r) spread in threshold voltage, since in a sub-100 nm channel only a limited and random number of dopants are implanted. In order to numerically model this problem, atomistic 3D simulations are required, using MC methods to randomly distribute the dopant atoms in the substrate. As the surface-potential variations induced by randomly distributed dopants are very similar to the ones induced by fixed oxide charge, it was only a small step to realize that this will also impact the RTS amplitude. Extensive modeling in this field has been performed by Asenov *et al* [55–59], showing that considering an atomistic doping approach fits experimental RTS amplitudes much better than assuming continuous doping (figure 3.26) [57]. In this modeling, only the number fluctuations have been considered, neglecting possible $\Delta\mu$ terms. This is related to the fact that it is not straightforward to implement a correct mobility model. It is clear that the atomistic model leads to higher RTS amplitudes, in line with the fixed-oxide-charge approach by Mueller *et al* [44].

Further calculation revealed that the impact of quantum effects is more pronounced in strong inversion, as there is less screening by the more distant mobile carrier centroid at lower currents [57]. In high inversion, screening is more effective, again reducing quantization effects. Moreover, the maximum RTS amplitude for a trap in the middle of the channel is fairly independent of the drain voltage (figure 3.27). The peaks shifts slightly towards the source for increasing V_{DS}, following the carrier density profile [57]. From the orders of magnitude difference in amplitude, one can readily explain the large scatter in this parameter, simply by considering a different spatial location of the oxide trap along the channel. This is related to the variation of the inversion-layer density n_s along the channel. It is obvious that for an oxide trap placed close to the source or drain contacts, small RTS amplitudes result, since the trapping of one electron will have only a minor

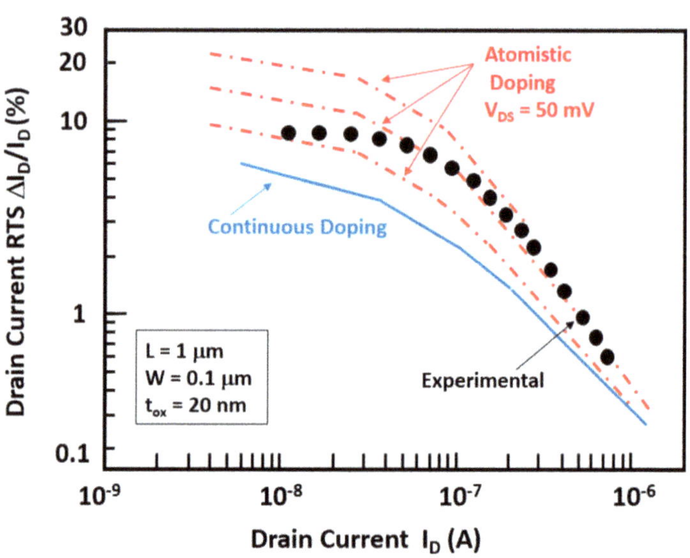

Figure 3.26. Dependence of RTS amplitude on the drain current in a relatively large MOSFET with channel width 0.1 μm, channel length 1 μm, oxide thickness 20 nm and channel doping 10^{17} cm^{-3} measured and simulated (for three different RDF configurations) at V_{DS} = 50 mV. (After Asenov *et al* [57].)

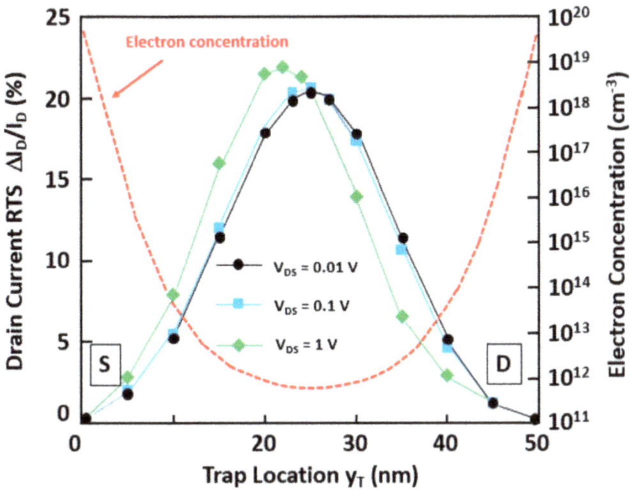

Figure 3.27. Dependence of the relative RTS amplitude on the position of the trapped charge along a line running from the source to the drain through the center of the channel of a 50 × 50 nm MOSFET at different drain voltages. The electron concentration at the interface along the channel is also presented. (After Asenov *et al* [57].)

impact there, assuming a simple number fluctuation model [57]. The simulations of figure 3.27 also demonstrate that an RTS in the source half of the transistor will increase with V_{DS}, while the opposite holds for a drain-half trap [57].

Using the concept of the trap length introduced in figure 3.18, one can calculate the relative RTS amplitude, as in figure 3.28, as a function of the effective device dimensions or as a function of the oxide thickness (figure 3.29) [57]. It is evident that the RTS amplitude increases for smaller-area devices but decreases with t_{ox}, due to the more efficient screening of the trapped charge by the gate [44, 50]. Finally, the role of current percolation in the surface-potential landscape induced by the RDFs is illustrated in figure 3.30, showing that much higher amplitudes can be obtained compared with a uniformly doped channel. Now, the highest RTS does not necessarily occur in the middle of the channel, but where the deepest valley in the potential landscape exists [57, 59].

This atomistic dopant approach has been extended to the case of 16 nm high-κ metal gate transistors [60] and to planar nano-MOSFETs [61]. It has been shown that transient and frequency domain modeling for RTS can be implemented in a 3D device simulator, including MC methods for random oxide traps and dopants [62].

A more analytical approach for estimating the impact of RDF on the RTS amplitude and the resulting change in threshold voltage has also been developed [63]. Based on a local cored-out area approach, the impact of a trap charge on the threshold voltage of a transistor can be expressed as:

$$\Delta V_T = \left(\frac{q}{C_{ox} W_{eff} L_{eff}}\right) \exp\left(\frac{q}{nkT}\left(V_T - V_{Tj}\right)\right) \tag{3.17}$$

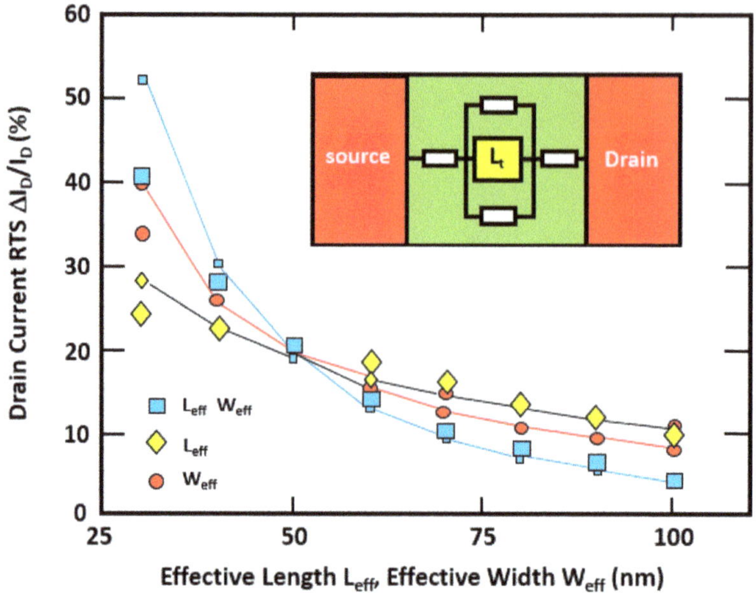

Figure. 3.28. Dependence of the relative RTS amplitude on the characteristic device dimensions for a set of square MOSFETs with fixed effective channel width, fixed effective channel length and with square geometry, at $V_{DS} = 10$ mV. Stand-alone symbols show numerical simulations; continuous lines show the analytical model of equation (3.12). (After Asenov *et al* [57].)

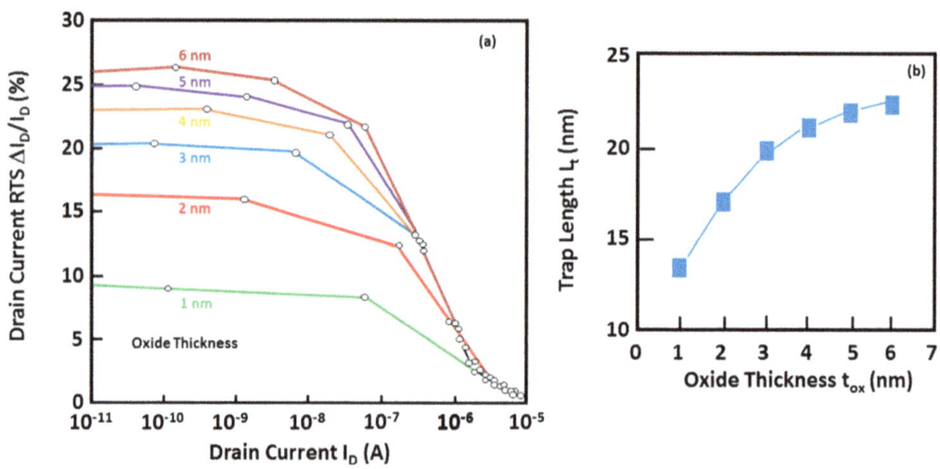

Figure 3.29. Dependence of the relative RTS amplitude on the drain current for a set of 50 × 50 nm MOSFETs with different oxide thickness and a single electron trapped in the middle of the channel (*a*). The trap depth in function of the oxide thickness is used as a fitting parameter (*b*). (After Asenov *et al* [57].)

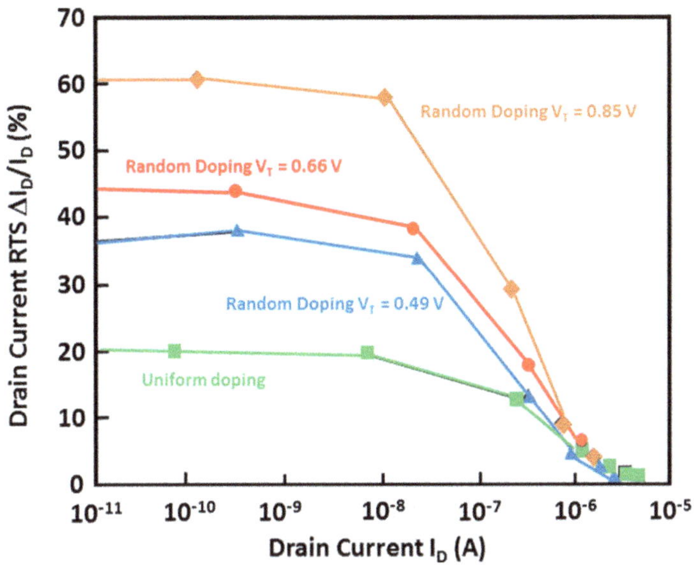

Figure 3.30. Dependence of the maximum RTS amplitudes on the drain current for three 50 × 50 nm devices with different atomistic doping. The corresponding dependence obtained from continuous doping simulations is also presented. (After Asenov *et al* [57].)

The model of equation (3.17) suggests that ΔV_T obeys not a normal distribution but a log-normal one, if the local threshold voltage V_{Tj} exhibits a normal distribution. The standard deviation of the local threshold voltage fluctuation is related to the surface-potential fluctuation as $\sigma_{V_{Tj}} = n\sigma_{\varphi_s}$. The standard deviation of $\log(\Delta V_T)$ is, therefore, given by [63]:

$$\sigma_{\log(\Delta V_T)} = \frac{q}{kT}\sigma_{\varphi_s} \tag{3.18}$$

The surface-potential fluctuation has been expressed by a superposition of Coulomb potentials created by random dopant atoms in the depletion layer, using Green's function for the electrostatic potential [63]. From this model, it was derived that the threshold shift induced by RTS will become a serious concern in 50 nm flash memories and beyond [63].

A 3D simulation of the RTS amplitude in accumulation- and inversion-mode trigate SOI MOSFETs has been reported by Yan *et al* [64]. It was concluded that the impact of RTS on the electrical parameters varies with the channel polarity and the physical location of the trapped electron. This is confirmed by the 3D statistical simulations of figure 3.31 [65–67], showing that significant relative amplitudes can be reached in SOI FFs, depending on the trap position in the oxide. $\Delta I_D/I_D$ decreases when EOT scales, due to screening by the metal gate electrode and amounts to ~$2.3 \times \Delta V_T$/SS, with ΔV_T the average threshold voltage shift induced by a single charge [67]. In addition, the relative RTS amplitude decreases with increasing temperature, since the subthreshold swing (SS) degrades with T (figure 3.32). A different gate voltage dependence can be observed for an oxide trap near the top

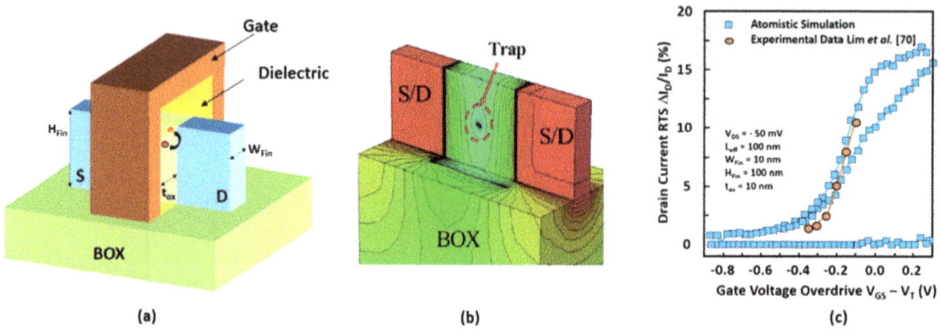

Figure 3.31. (*a*) Schematic of trapping/detrapping of carriers in an SOI FF and (*b*) the perturbed potential contour by the trapped charge. (*c*) Calibration of RTS amplitude for an SOI FF with L_{eff}~100 nm, W_{Fin}~10 nm and t_{ox}~10 nm. The experimental data from Lim *et al* [70] fall within the range of the simulations with three possible trap locations. (After Fan *et al* [67].)

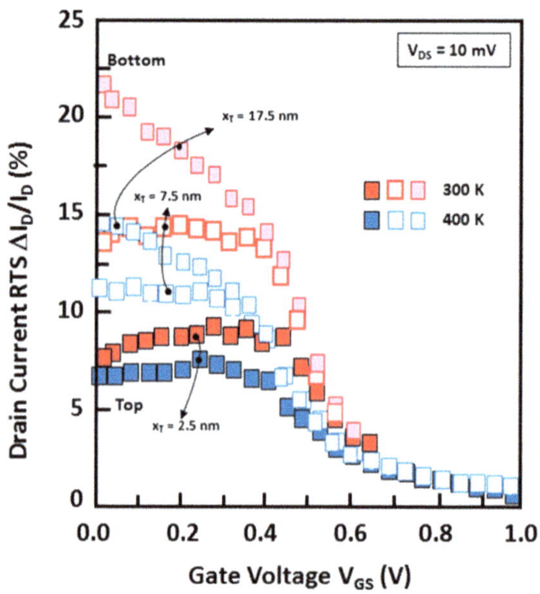

Figure 3.32. Temperature dependence of RTS amplitude at $T = 300$ K and $T = 400$ K. The trap is placed at $0.5L_{\text{eff}}$ with four different locations (*z*-direction) from bottom to top of the silicon fin. EOT = 0.65 nm. (After Fan *et al* [67].)

or the bottom of the fin; the difference stems from the change in the charge-carrier distribution [67]. It is clear from figure 3.32 that a trap at the bottom of the fin has a stronger impact than one at the top of the fin.

In the case of the independent-gate mode, the RTS amplitude can be much higher due to the poorer electrostatics (figure 3.33). At the same time, it has been shown that planar bulk MOSFETs can reach much higher $\Delta I_{\text{D}}/I_{\text{D}}$ than in undoped FFs due to the amplification effect by the random dopants [67]. A detailed comparison based

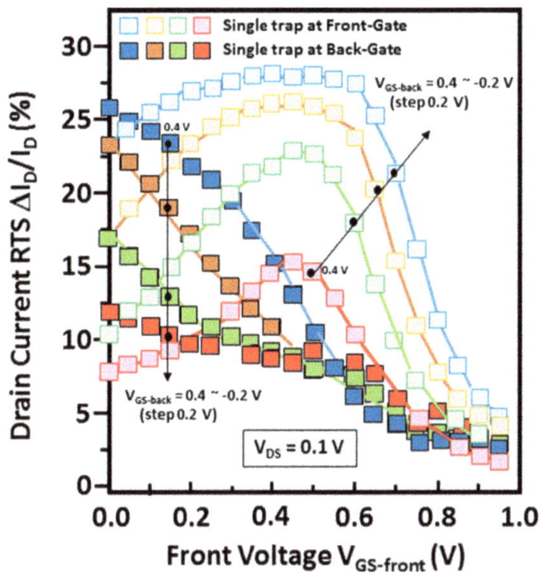

Figure 3.33. Dependence of RTS amplitude on $V_{GS\text{-front}}$ at various $V_{GS\text{-back}}$ biases for SOI FF operation in independent-gate mode. The trap is placed either at the center of the front-gate or the back-gate sidewall interfaces for a FF with EOT = 0.65 nm. (After Fan *et al* [67].)

on atomistic simulations of bulk planar, fully depleted SOI (FDSOI) and FFs has revealed that the RTS-induced variability is ranked in descending order: bulk > FDSOI > FF, which is associated with the higher channel doping in bulk compared with the other technologies [68, 69].

3.5 Novel measurement and analysis methods

Acknowledging that RTS is a source of dynamic device and circuit variability implies that a statistical assessment of its parameter (time constants, amplitude, activation energy, etc) distributions has to be carried out in order to estimate the impact on the real-time operation and reliability of logic, analog, and memory components and circuits. This means that many devices have to be measured for a sufficiently long time, rendering a full RTS characterization based on standard time-domain measurements on individual components a time and measurement resource consuming exercise. An important question regarding RTS time-domain measurements is: which sampling rate is adequate to determine the time constants with sufficient accuracy? As a rule of thumb, one can state that in order to reach an accuracy of 10% for the time constants, at least 200 transitions are required and a sampling rate of at least 100 times the average up and down time [71]. A more rigorous analysis based on MC simulations has pointed out that, depending on the τ_e/τ_c ratio, the required sampling rate may vary from 70 times the original τ_e for $\tau_e/\tau_c = 1$ to 1000 times when $\tau_e/\tau_c \sim 100$ [72]. All in all, this means that accurate RTS measurements become quite lengthy, in particular if slow signals occur.

Several methods have been developed to speed up the measurements without compromising too much on accuracy. An elegant solution is offered by the so-called *fast ramped voltage characterization* of the impact of a single trap on the V_T variability [73]. The RTS data are collected during a ramped I_D–V_{GS} sweep, which at the same time accesses the gate voltage dependence of the time constants. An example is shown in figure 3.34, where ΔI_D can be transformed into a ΔV_T using the transconductance. The magnitude of ΔV_T is approximately constant, indicating a single trap origin. In this way, the whole V_{GS} range from weak to strong inversion is scanned for the presence of active RTS, i.e. with $\tau_c \sim \tau_e$, in a reasonable amount of time.

Employing this method on 28 nm replacement metal gate (RMG) n- and pMOSFETs reveals that on average between 0.5 and 1 active oxide traps are found per device, with a higher average amplitude for p- compared with nMOS due to the larger average distance from the interface in the latter case [73]. The ΔV_T follows an exponential distribution, as will be shown more extensively in the following

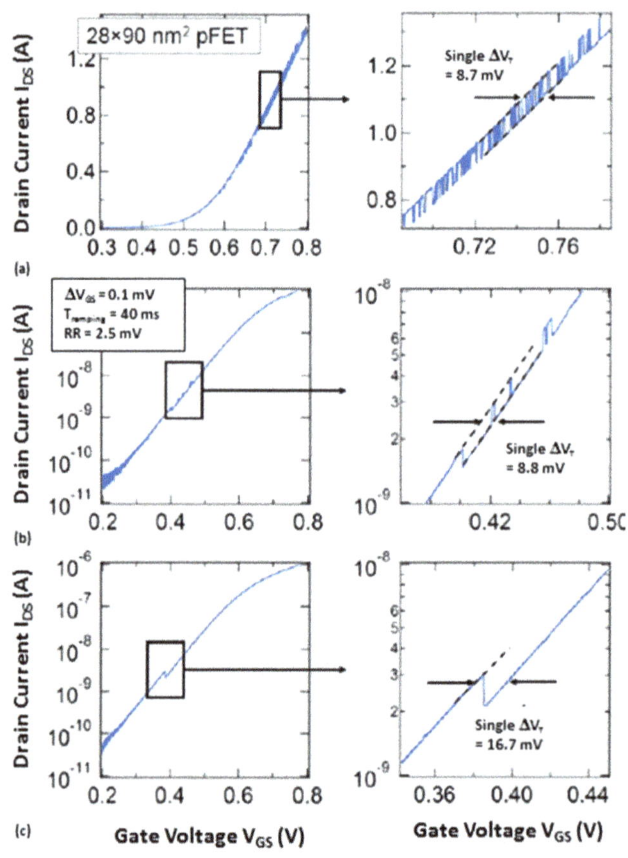

Figure 3.34. RTS features are visible in the I_D–V_{GS} curves in (*a*) linear, and (*b*) and (*c*) subthreshold regimes. The ΔV_T caused by a single trap (single ΔV_T) is to a first approximation independent of V_{GS}, but differs from trap to trap. (Reproduced with permission from Toledano-Luque *et al* [73]. Copyright 2014 IEEE.)

paragraph. In order to also derive the gate voltage dependence of the time constants, one assumes that τ_c and τ_e exhibit an exponential dependence on the gate voltage around the crossing point where $\tau_c = \tau_e$ [73]. In other words, one considers

$$\tau_c = \tau_0 \exp\left[\alpha_c(V_{GS} - V_{GS0})\right] \quad (3.19a)$$

$$\tau_e = \tau_0 \exp\left[\alpha_e(V_{GS} - V_{GS0})\right] \quad (3.19b)$$

with τ_0 and V_{GS0} the time constant and the gate bias corresponding to the crossing point, and α_c and α_e the respective coupling factors. The τ_c and τ_e versus V_{GS} dependence is then found by applying a novel maximum likelihood method [73].

Based on the statistical model, the coupling factors and the crossing point can be estimated from a given set of data containing the state transition gate voltages and the dwelling times coming from the I_D–V_{GS} traces, provided that τ_0 and V_{GS0} are within certain limits. The experimental time window for a ramp rate of 40 ms is between 0.2 s and 10 s. The mark space ratio is limited as in the conventional RTS technique. For the majority of traps, one finds that $|\alpha_e| < |\alpha_c|$. This implies that a smaller gate voltage swing is required to vary τ_c over an order of magnitude. To a first approximation, the following relationship holds [73]

$$\frac{x_T}{t_{ox}} = -\frac{kT}{q}\frac{\partial \ln\left(\frac{\tau_c}{\tau_e}\right)}{\partial V_{GS}} = -\frac{kT}{q}(\alpha_c - \alpha_e) \quad (3.20)$$

so that the trap depth can be directly derived as well. As shown in figure 3.35, traps in p–channel transistors are closer to the interface than in nMOSFETs, explaining

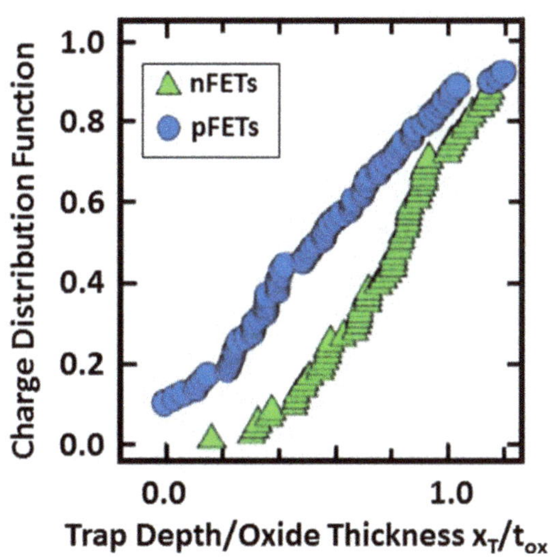

Figure 3.35. Trap depths x_t extracted from the coupling factors. Traps in nFETs are located further away from the Si/SiO_2 interface compared to pFET traps. (Reproduced with permission from Toledano-Luque *et al* [73]. Copyright 2014 IEEE.)

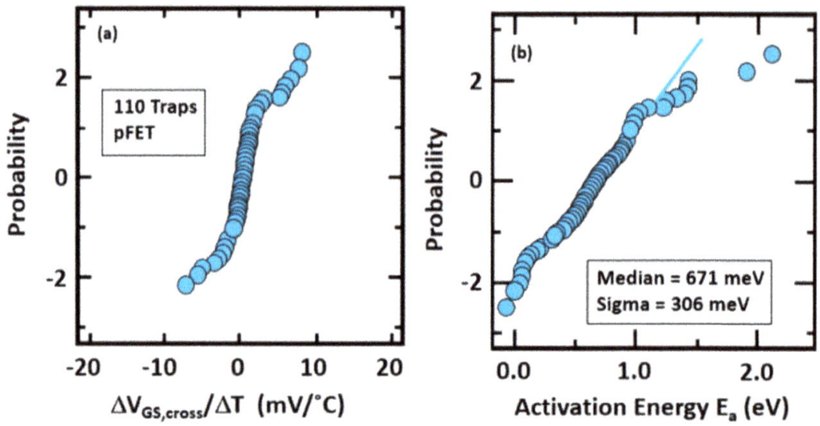

Figure 3.36. A low sensitivity of V_{GS0} to temperature (*a*) but a wide distribution of activation energy obtained from more than 100 traps from about 400 pFETs. (Reproduced with permission from Toledano-Luque *et al* [73]. Copyright 2014 IEEE.)

their larger ΔV_T. In addition, the product of $(\alpha_c - \alpha_e) \times t_{ox}$ stays approximately constant for thicker oxide pMOSFETs [73].

According to figure 3.36, V_{GS0} exhibits limited T dependence, whereby both an increase or a decrease with T can occur [73]. A single ΔV_T shift is also a little temperature-dependent. The median value for E_a from an Arrhenius plot of τ_0 is 700 meV for the studied devices. It was also shown that at higher temperature, more traps become active due to the shorter τ_c, which among others also leads to a stronger NBTI (see chapter 5 for more details).

Other methods consider alternative bias pulsing schemes to address the RTS time constants, such as, e.g., the incremental trap response technique proposed recently [74]. In contrast, designing special test structures where large amounts of transistors are combined in a memory-like array enables one to collect, for example, the statistical distribution of ΔV_T shifts in a short period of time. This will be discussed and illustrated in the next chapter.

Some dedicated time-domain methods have recently been proposed for a high-resolution analysis of RTSs [75] and multiple-trap signals [76, 77] in high-κ gate stacks. The presence of the SiO_2 IL can give rise to background or 'ghost' noise which affects the accurate extraction of the RTS amplitude. Assuming that this ghost noise obeys Gaussian statistics, a hidden Markov method can be employed to extract the RTS parameters more accurately [75]. Another helpful tool is the use of so-called TLPs, as illustrated by figure 3.37, where the drain current at time interval $i + 1$ is represented versus I_D at time i [76, 77]. A TLP is a graphical data analysis technique for determining whether an autocorrelation structure exists within a time series. If the current amplitude sampled at two discrete times (t_i and t_{i+1}) does not change, this results in a data point along the main diagonal of the TLP with x- and y-coordinates corresponding to the discrete current levels. Use of a TLP is particularly interesting if multiple traps are present at the same time, enabling a

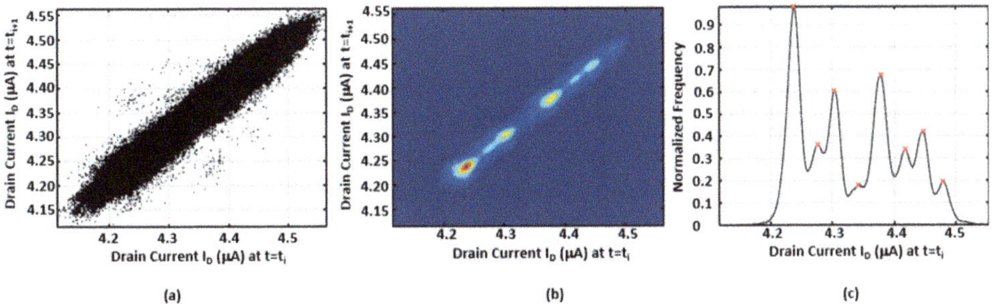

Figure 3.37. Comparison between (*a*) standard and (*b*) enhanced TLP for a triple-trap RTS. (*c*) Enhanced TLP diagonal along with detected RTS levels. (Reproduced with permission from Realov and Sheppard [76]. Copyright 2013 IEEE.)

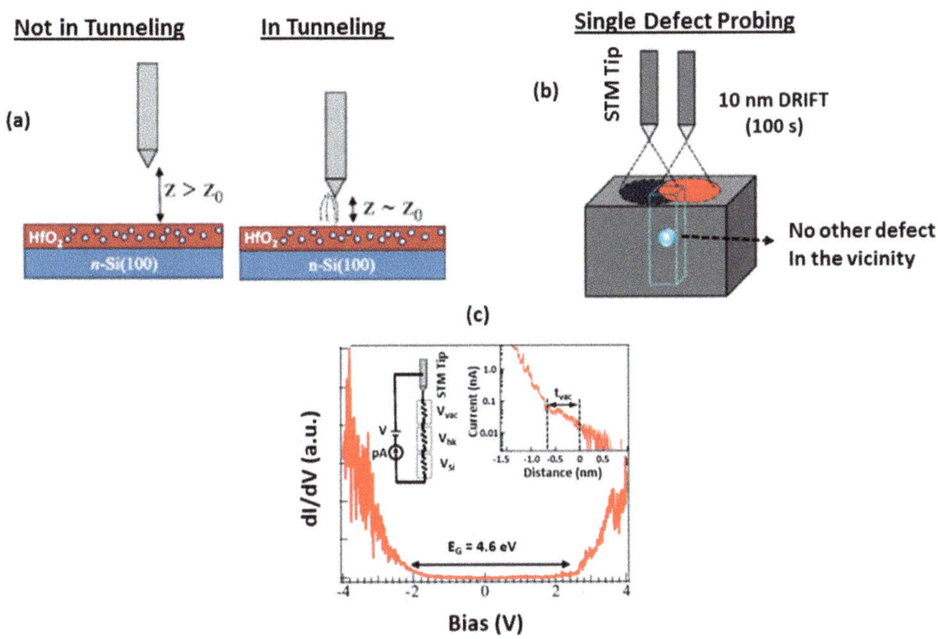

Figure 3.38. (*a*) Schematic diagram of the STM configuration. Only when the tip is at a distance of ($z < z_0$) is the tunneling condition satisfied. The voltage drop experienced by the dielectric is a fraction of the actual voltage applied to the tip. (*b*) Illustration of the possibility of detecting RTS from the same single defect with no other defect in the vicinity in spite of the inherent drift of 0.1 nm s^{-1} in the STM set-up. (*c*) Energy gap of $E_G = 4.6$ eV, as measured using the lock-in technique ($f_0 = 2.7$ kHz and $\Delta V = 10$ mV). The inset shows a simple equivalent circuit diagram of the experiment, where V_{vac}, V_{HK} and V_{Si} represent the voltage drop across the tunnel gap, high-κ and Si substrate, respectively. The I–z curve (also shown as an inset) is measured on a HfO_2 island (grain) to determine the vacuum gap separation (t_{vac}) used in the voltage drop calculation. (Reproduced with permission from Thamankar *et al* [81]. Copyright 2016 AIP Publishing.)

better separation of the different trap parameters. This method can be further improved by using weighted or enhanced representations [76, 78, 79].

While scanning tunneling microscopy (STM), depicted in figure 3.38, offers potentially a high spatial resolution on the order of ~10 nm to study tunneling traps in a gate oxide, few results have been reported. Early reports showed the feasibility in the Si/SiO_2 system [80]. Quite recently [81], the first STM study of process-induced RTS in a HfO_2 gate dielectric was presented. These signals have been ascribed to oxygen vacancies in the gate stack [81]. Both two-level and complex, multi-level RTSs associated with a defect cluster, were identified. It was shown that the time constants follow a bimodal distribution and not a single exponential, which evidences the existence of a metastable defect state [81]. A hidden Markov model is necessary to reveal these fine details of the studied switching events. One of the limitations of the method is the spatial drift of the probe tip on the order of 0.1 nm s^{-1}, which only allows reliable detection of 'fast' traps with time constants of a few milliseconds up to a few tens of seconds [81].

3.6 *Ab initio* modeling of RTS in gate dielectrics

One important question which still remains to be answered is: what is the microscopic origin of the oxide defects responsible for RTS (and BTI, see chapter 5)? A notorious suspect, well-known from electron paramagnetic resonance (EPR) studies are the well-known oxygen vacancies (E' centers). In order to assess whether these defects are really responsible and can explain the behavior of RTS defects, derived from time-domain and BTI measurements, first principles *ab initio* calculations have become an indispensable tool. Generally, DFT calculations are performed assuming certain defect configurations, as shown in figure 3.39 [82]. These calculations allow

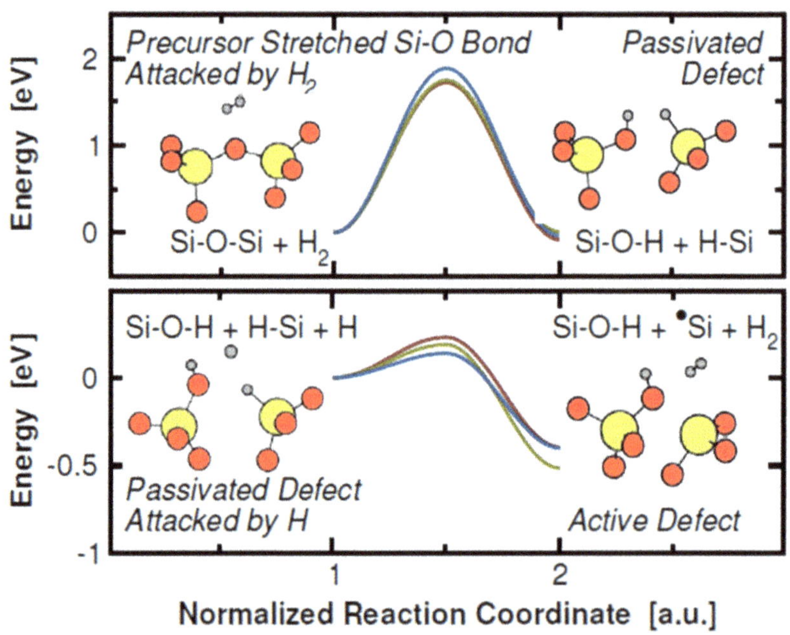

Figure 3.39. Possible creation scenario of the hydroxyl E' center during high-temperature processing via the attack of a stretched Si–O–Si bond by H_2, which is later attacked by H to eventually form the active defect. Creation is limited by the precursor density, not the H concentration. (Reproduced with permission from Grasser *et al* [82]. Copyright 2014 IEEE.)

one to estimate the energy levels in the oxide band gap and the reaction barriers between different (meta)stable configurations, depending on the charge state of the defect. This method also points out possible reaction paths upon charge capture. In addition, for amorphous materials it is important to address the statistical spread in defect parameters, which is in principle also possible using a DFT scheme.

Recently, several groups have started such *ab initio* calculations on defects in SiO_2 [82, 83] or in HfO_2 [84]. One of the main conclusions is that the oxygen vacancy has unfavorable energy barriers to fit the experimental data (figure 3.40). However, SiO_2 hydroxyl E' centers, represented in figure 3.40, are more likely candidates to explain the RTS behavior. The involvement of the highly mobile hydrogen atom (present in high concentrations) possibly explains the frequently observed disappearance and re-appearance of oxide-trap related RTS, also called anomalous RTS [85]. On the other hand, similar studies for HfO_2 gate dielectrics indicate double oxygen vacancies as candidate RTS centers [84]. DFT calculations have also been used to investigate the so-called volatility of the oxide traps [86], where RTS can disappear and re-appear on a wide time scale [84]. It is believed that a large fraction of RTS traps in fact exhibit such behavior (up to 60%) [86]. As this turns out to be a normal (and not an anomalous) feature of RTS and BTI traps, the origin of the phenomenon needs to be understood more clearly. Again, hydrogen seems to play an important role in oxide defect volatility, transforming hydroxyl E' into an active or an inactive state with time [86].

As these are only initial results, much more work needs to be done in order to firmly identify the defects responsible for RTS and BTI. Such identification is of

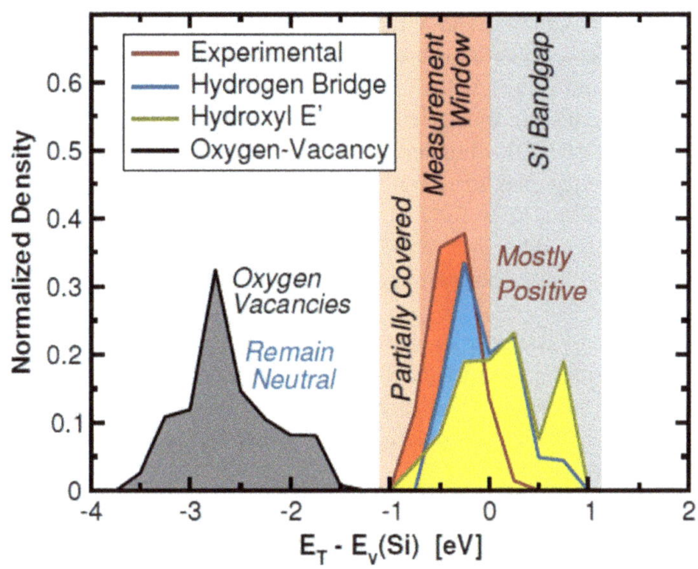

Figure 3.40. The fundamental parameter which decides which trap can be charged for a certain stress (V_{GS}^{H}) and recovery (V_{GS}^{L}) combination is the energy level E_T. Clearly, the oxygen vacancy/E' center is too low in energy, while both the hydrogen bridge and the hydroxyl E' center are in good agreement with the data inside the experimental window. (Reproduced with permission from Grasser *et al* [82]. Copyright 2014 IEEE.)

crucial importance to mitigate these phenomena in deep submicron and nanometer devices and circuits.

References

[1] Palma A, Godoy A, Jiménez-Tejada J A, Carceller J E and López-Villanueva J A 1997 Quantum two-dimensional calculation of time constant of random telegraph signals in metal–oxide–semiconductor structures *Phys. Rev.* B **56** 9565–74

[2] Lukyanchikova N B, Petrichuk M V, Garbar N P, Simoen E and Claeys C 2000 Influence of substrate voltage on the random telegraph signal parameters in submicron n-channel metal–oxide–semiconductor field-effect transistors under a constant inversion charge density *Appl. Phys.* A **70** 345–53

[3] Çelik-Butler Z and Wang F 2000 Effects of quantization on random telegraph signals observed in deep-submicron MOSFETs *Microelectron. Reliab.* **40** 1823–31

[4] Çelik-Butler Z 2002 Low-frequency noise in deep submicron metal–oxide–semiconductor field-effect transistors *IEE Proc.—Circuits Devices Syst.* **149** 23–31

[5] Lu M-P and Chen M-J 2005 Oxide-trap-enhanced Coulomb energy in a metal–oxide–semiconductor system *Phys. Rev.* B **72** 235417

[6] Amarasinghe N V, Çelik-Butler Z and Keshavarz A 2001 Extraction of oxide trap properties using temperature dependence of random telegraph signals in submicron metal–oxide–semiconductor field-effect transistors *J. Appl. Phys.* **89** 5526–32

[7] Amarasinghe N V, Çelik-Butler Z, Zlotnicka A and Wang F 2003 Model for random telegraph signals in sub-micron MOSFETs *Solid-State Electron.* **47** 1443–9

[8] Pud S, Gasparyan F, Petrychuk M, Li J, Offenhäusser A and Vitusevich S A 2015 Single trap dynamics in electrolyte-gated Si-nanowire field effect transistor *J. Appl. Phys.* **115** 233705

[9] Gasparyan F, Zadorozhnyi I and Vitusevich S 2015 Single trap in liquid gated nanowire FETs: capture time behavior as a function of current *J. Appl. Phys.* **117** 1–5

[10] Yoo S-W, Son Y and Shin H 2013 Capture cross section of traps causing random telegraph noise in gate-induced drain leakage current *IEEE Trans. Electron Devices* **60** 1268–71

[11] Tu K-C and Chen M-J 2012 Random telegraph signals under external mechanical stress: a new method to probe trap structural relaxation in MOSFET gate dielectrics *Extended Abstract of the 2012 Int. Conf. Solid State Devices and Materials* (Kyoto, Japan) pp 837–8

[12] Longoni A, Gatti E and Sacco R 1995 Trapping noise in semiconductor devices: a method for determining the noise spectrum as a function of the trap position *J. Appl. Phys.* **78** 6283–97

[13] Amarasinghe N V, Çelik-Butler Z and Vasina P 2000 Characterization of oxide traps in 0.15 μm^2 MOSFETs using random telegraph signals *Microelectron. Reliab.* **40** 1875–81

[14] Çelik-Butler Z, Vasina P and Amarasinghe N V 2000 A method for locating the position of oxide traps responsible for random telegraph signals in submicron MOSFETs *IEEE Trans. Electron Devices* **47** 646–8

[15] Yang S, Lee H and Shin H 2007 Simultaneous extraction of locations and energies of two independent traps in gate oxide from four-level RTS noise *Extended Abstract of the 2007 Int. Conf. on Solid State Devices and Materials* pp 900–1

[16] Yang S, Lee H and Shin H 2008 Simultaneous extraction of locations and energies of two independent traps in gate oxide from four-level random telegraph signal noise *Jpn. J. Appl. Phys.* **47** 2606–9

[17] Kang D, Kim J, Lee D, Park B-G, Lee J D and Shin H 2009 Extraction of vertical, lateral locations and energies of hot-electrons-induced traps through the random telegraph noise *Jpn. J. Appl. Phys.* **48** 04C034

[18] Park S, Lee S, Kang Y, Park B-G, Lee J-H, Lee J, Jin G and Shin H 2010 Extracting accurate position and energy level of oxide trap generating random telegraph noise (RTN) in recessed channel MOSFET's *Tech. Dig. Int. Electron Devices Meeting—IEDM10* (New York: IEEE) pp 337–40

[19] Nagumo T, Takeuchi K, Hase T and Hayashi Y 2010 Statistical characterization of trap position, energy, amplitude and time constants by RTN measurement of multiple individual traps *Tech. Dig. Int. Electron Devices Meeting—IEDM10* (New York: IEEE) pp 628–31

[20] Svedlakova V, Sikula J, Chvatal M, Pavelka J, Tacano M and Toita M 2012 Noise in submicron metal–oxide–semiconductor field effect transistors: lateral electron density distribution and active trap position *Jpn. J. Appl. Phys.* **51** 1–5

[21] Chen C-E *et al* 2013 Investigation of lateral trap position by random telegraph signal analysis in moderate inversion in n-channel MOSFETs *ECS Solid State Lett.* **2** Q90–Q92

[22] Chen C-E *et al* 2015 A method to determine the located region of lateral trap position by analysis of three-level random telegraph signals in n-MOSFETs *ECS Solid State Lett.* **4** Q47–Q49

[23] Chang C M, Chung S S, Hsieh Y S, Cheng L W, Tsai C T, Ma G H, Chien S C and Chun S W 2008 The observation of trapping and detrapping effects in high-*k* gate dielectric MOSFETs by a new gate current random telegraph noise (I_G-RTN) approach *Tech. Dig. Int. Electron Devices Meeting—IEDM08* (New York: IEEE)

[24] Lee S, Cho H-J, Son Y, Lee D S and Shin H 2009 Characterization of oxide traps leading to RTN in high-*k* and metal gate MOSFETs *Tech. Dig. Int. Electron Devices Meeting—IEDM09* (New York: IEEE) pp 763–6

[25] Cho H-J, Lee S, Park B-G and Shin H 2010 Extraction of trap energy and location from random telegraph noise in gate leakage current (I_g RTN) of metal-oxide semiconductor field effect transistor (MOSFET) *Solid-State Electron.* **54** 362–7

[26] Cho H-J, Son Y, Oh B, Lee S, Lee J-H, Park B-G and Shin H 2010 Observation of slow oxide traps at MOSFETs having metal/high-*k* gate dielectric stack in accumulation mode *IEEE Trans. Electron Devices* **57** 2697–703

[27] Cho H-J, Son Y, Oh B-C, Lee S, Lee J-H, Park B-G and Shin H 2010 Study on time constants of random telegraph noise in gate leakage current through hot-carrier stress test *IEEE Electron Device Lett.* **31** 1029–31

[28] Cho H-J, Son Y, Jang S and Shin H 2011 Study of slow oxide trap creating random telegraph noise within a gate edge overlap region in inversion mode *Appl. Phys. Lett.* **88** 1–3

[29] Lee J-W, Shin H and Lee J-H 2010 Characterization of random telegraph noise in gate induced drain leakage current of n- and p-type metal–oxide–semiconductor field-effect transistors *Appl. Phys. Lett.* **96** 043502

[30] Lee J-W, Lee B H, Shin H and Lee J-H 2010 Investigation of random telegraph noise in gate-induced drain leakage and gate edge direct tunneling currents of high-*k* MOSFETs *IEEE Trans. Electron Devices* **57** 913–8

[31] Lee J-W, Park C H, Shin H, Park B-G and Lee J-H 2011 Accurate extraction of ΔI/I due to random telegraph noise in gate edge current of high-*k* n-type metal–oxide–semiconductor field-effect transistors under accumulation mode *Appl. Phys. Lett.* **98** 023505

[32] Lee J-W and Lee J-H 2012 Model of random telegraph noise in gate-induced drain leakage current of high-*k* gate dielectric metal–oxide–semiconductor field-effect transistors *Appl. Phys. Lett.* **100** 1–3

[33] Cho H-J, Son Y, Oh B, Jang S, Lee J-H, Park B-G and Shin H 2011 Investigation of gate etch damage at metal/high-*k* gate dielectric stack through random telegraph noise in gate edge direct tunneling current *IEEE Electron. Device Lett.* **32** 569–71

[34] Gia Q N, Yoo S-W, Lee H and Shin H 2014 Dependence on an oxide trap's location of random telegraph noise (RTN) in GIDL current of n-MOSFET *Solid-State Electron.* **92** 20–3

[35] Shin J, Jeon S, Kim H S and Yoo S-W 2015 Analysis on the variable junction leakage in MOS transistors due to interaction between two traps *Jpn. J. Appl. Phys.* **54** 41044

[36] Yoo S-W, Shin J, Seo Y, Kim H, Jeon S, Kim H and Shin H 2015 Characterizing traps causing random telegraph noise during trap-assisted tunneling gate-induced drain leakage *Solid-State Electron.* **109** 42–6

[37] Castellani N, Monzio Compagnoni C, Mauri A, Spinelli A S and Lacaita A L 2012 Three-dimensional electrostatics—and atomistic doping-induced variability of RTN time constants in nanoscale MOS devices. Part I: Physical investigation *IEEE Trans. Electron Devices* **59** 2488–94

[38] Monzio Compagnoni C, Castellani N, Mauri A, Spinelli A S and Lacaita A L 2012 Three-dimensional electrostatics—and atomistic doping-induced variability of RTN time constants. Part II: Spectroscopic implications *IEEE Trans. Electron Devices* **59** 2495–500

[39] Adamu-Lema F, Monzio Compagnoni C, Amoroso S M, Castellani N, Gerrer L, Markov S, Spinelli A S, Lacaita A L and Asenov A 2013 Accuracy and issue of the spectroscopic analysis of RTN traps in nanoscale MOSFETs *IEEE Trans. Electron Devices* **60** 833–9

[40] Abe K, Teramoto A, Sugawa S and Ohmi T 2011 Understanding of traps causing random telegraph noise based on experimentally extracted time constants and amplitude *Proc. Annual Int. Reliability Physics Symp.—IRPS11* (New York: IEEE) pp 381–6

[41] Illarionov Y Y, Bina M, Tyaginov S E, Rott K, Reisinger H, Kaczer B and Grasser T 2014 A reliable method for the extraction of the lateral position of defects in ultra-scaled MOSFETs *Proc. Annual Int. Reliability Physics Symp.—IRPS14* (New York: IEEE) pp XT13.1–13.6

[42] Illarionov Y Y, Bina M, Tyaginov S E and Grasser T 2014 An analytical approach for the determination of the lateral trap position in ultra-scaled MOSFETs *Jpn. J. Appl. Phys.* **53** 1–4

[43] Illarionov Y, Bina M, Tyaginov S, Rott K, Kaczer B, Reisinger H and Grasser T 2015 Extraction of the lateral position of border traps in nanoscale MOSFETs *IEEE Trans. Electron Devices* **62** 2730–7

[44] Mueller H H and Schulz M 1996 Conductance modulation of submicrometer metal–oxide–semiconductor field-effect transistors by single-electron trapping *J. Appl. Phys.* **79** 4178–86

[45] Campbell J P, Qin J, Cheung K P, Yu L C, Suehle J S, Oates A and Sheng K 2009 Random telegraph noise in highly scaled nMOSFETs *Proc. 47th Annual Int. Reliability Physics Symp.—IRPS09* (New York: IEEE) pp 382–8

[46] Schulz M 1993 Coulomb energy of traps in semiconductor space-charge regions *J. Appl. Phys.* **74** 2649–57

[47] Simoen E, Dierickx B, Claeys C L and Declerck G J 1992 Explaining the amplitude of RTS noise in submicrometer MOSFET's *IEEE Trans. Electron Devices* **39** 422–9

[48] Mueller H H, Schirl U and Schulz M 1995 Conductance modulation by single-electron trapping in sub-μm MOSFETs *Microelectron. Eng.* **28** 163–6

[49] Chen M-J, Lee C-C and Lu M-P 2008 Probing a nonuniform two-dimensional electron gas with random telegraph signals *J. Appl. Phys.* **103** 034511

[50] Zhang P, Zhuang Y Q, Ma Z F, Bao L, Du L and Bao J L 2008 The impact of gate-image charge on RTS amplitudes in ultra-thin gate oxide n-MOSFETs *Semicond. Sci. Technol.* **23** 125037

[51] Mueller H H and Schulz M 1997 Statistical evaluation of random telegraph signal amplitudes in sub-μm MOSFETs *Microelectron. Reliab.* **36** 223–6

[52] Forbes L and Miller D A 2008 A percolation model for random telegraph signals in metal–oxide–silicon field effect transistor drain current *Appl. Phys. Lett.* **93** 1–3

[53] Chen M-J, Tu K-C, Wang H-H, Chen C-L, Lai S-Y and Liu Y-S 2014 A statistical model for the headed and tail distributions of random telegraph signal magnitudes in nanoscale MOSFETs *IEEE Trans. Electron Devices* **61** 2495–502

[54] Chen M-J, Tu K-C, Chuang L-Y and Wang H-H 2015 Graphically transforming Mueller–Schulz percolation criteria to random telegraph signal magnitudes in scaled FETs *IEEE Electron Device Lett.* **36** 217–9

[55] Asenov A, Balasubramanian R, Brown A R, Davies J H and Saini S 2000 Random telegraph signal amplitudes in sub 100 nm (decanano) MOSFETS: a 3D 'atomistic' simulation study *Tech. Dig. Int. Electron Devices Meeting—IEDM* (New York: IEEE) pp 279–82

[56] Asenov A, Balasubramanian R, Brown A R and Davies J H 2000 Effect of single-electron interface trapping in decanano MOSFETs: a 3D atomistic simulation study *Superlattices Microstruct.* **27** 411–6

[57] Asenov A, Balasubramanian R, Brown A R and Davies J H 2003 RTS amplitudes in decananometer MOSFETs: 3D simulation study *IEEE Trans. Electron Devices* **50** 839–45

[58] Alexander C L, Brown A R, Watling J R and Asenov A 2005 Impact of single charge trapping in nano-MOSFETs—electrostatics versus transport effects *IEEE Trans. Nanotechnol.* **4** 339–44

[59] Bukhori M F, Roy S and Asenov A 2008 Statistical aspects of reliability in bulk MOSFETs with multiple defect states and random discrete dopants *Microelectron. Reliab.* **48** 1549–52

[60] Li Y and Chen H-W 2012 Statistical device simulation of physical and electrical characteristic fluctuations in 16-nm-gate high-*k*/metal gate MOSFETs in the presence of random discrete dopants and random interface traps *Solid-State Electron.* **77** 12–9

[61] Zhongfa M, Peng Z, Yong W, Weihua L, Yiqi Z and Lei D 2010 A percolation study of RTS noise amplitudes in nano-MOSFETs by Monte Carlo simulation *Microelectron. Reliab.* **50** 179–82

[62] Higashi Y, Momo N, Momose H S, Ohguro T and Matsuzawa K 2011 Comprehensive understanding of random telegraph noise with physics based simulation,*Symp. on VLSI Technol. Dig. Tech. Papers*, pp 200–201

[63] Sonoda K, Ishikawa K, Eimori T and Tsuchiya O 2007 Discrete dopant effect on statistical variation of random telegraph signals *IEEE Trans. Electron Devices* **54** 1918–25

[64] Yan R, Cullen A, Afzalian A, Ferain I, Lee C-W, Dehdashti N, Razavi R and Colinge J P 2010 3D simulation of RTS amplitude in accumulation-mode and inversion-mode trigate SOI MOSFETs *Proc. EUROSOI 2010*, Grenoble, France, pp 89–90

[65] Fan M-L, Hu V P-H, Chen Y-N, Su P and Chuang C-T 2011 Impacts of single trap induced random telegraph noise on FinFET devices and SRAM cell stability *Proc. 2011 IEEE Int. SOI Conf.* (New York: IEEE) pp 1–2

[66] Fan M-L, Hu V P-H, Chen Y-N, Su P and Chuang C-T 2012 Impacts of random telegraph noise on FinFET devices, 6 T SRAM cell, and logic circuits *Proc. Annual Int. Rel. Phys. Symposium—IRPS12*, pp CR1.1–CR.1.6

[67] Fan M-L, Hu V P-H, Chen Y-N, Su P and Chuang C-T 2012 Analysis of single-trap-induced random telegraph noise on FinFET devices, 6 T SRAM cell, and logic circuits *IEEE Trans. Electron Devices* **59** 2227–34

[68] Gerrer L, Amoroso S M, Hussin R and Asenov A 2014 A comparative study of variability of RTN power spectral densities in bulk and SOI MOSFETs *Proc. EUROSOI* pp 1–2

[69] Gerrer L, Amoroso S M, Hussin R and Asenov A 2014 RTN distribution comparison for bulk, FDSOI and FinFET devices *Microelectron. Reliab.* **54** 1749–52

[70] Lim Y F *et al* 2006 Random telegraph signal noise in gate-all-around Si-FinFET with ultra-narrow body *IEEE Electron Device Lett.* **27** 765–8

[71] Kirton M J and Uren M J 1989 Noise in solid-state microstructures: a new perspective on individual defects, interface states and low-frequency (1/*f*) noise *Adv. Phys.* **38** 367–468

[72] Kaplia G and Reddy V 2014 Impact of sampling rate on RTN time constant extraction and its implications on bias dependence and trap spectroscopy *IEEE Trans. on Device and Materials Reliability* **14** 616–22

[73] Toledano-Luque M, Degraeve R, Roussel P J, Ragnarsson L-Å, Chiarella T, Horiguchi N, Mocuta A and Thean A 2014 Fast ramped voltage characterization of single trap bias and temperature impact on time-dependent V_{TH} variability *IEEE Trans. Electron Devices* **61** 3139–44

[74] Guo S, Ren P, Wang R, Yu Z, Luo M, Zhang X and Huang R 2014 A new efficient method for characterizing time constants of switching oxide traps *Proc. Annual Int. Reliability Physics Symp.—IRPS14* (New York: IEEE) pp XT.14.1–14.4

[75] Li M, Wang R, Zou J and Huang R 2013 Characterization of random telegraph noise in scaled high-*κ*/metal gate MOSFETs with SiO_2/HfO_2 gate dielectrics *ECS Trans.* **52** 941–6

[76] Realov S and Sheppard K L 2013 Analysis of random telegraph noise in 45-nm CMOS using on-chip characterization system *IEEE Trans. Electron Devices* **60** 1716–22

[77] Tsuchiya T, Tamura N, Sakakidani A, Sonoda K, Kamei M, Yamakawa S and Kuwabara S 2013 Characterization of oxide traps participating in random telegraph noise using charging history effects in nano-scaled MOSFETs *ECS Trans.* **58** 265–79

[78] Martin-Martinez J, Diaz J, Rodriguez R, Nafria M and Aymerich X 2014 New weighted time lag plot method for the analysis of random telegraph signals *IEEE Electron Device Lett.* **35** 479–81

[79] Simoen E, Fang W, Luo J, Zhao C and Claeys C 2016 Random telegraph noise: the key to single defect studies in nano-devices *Thin Solid Films* **613** 2–5

[80] Welland M E and Koch R H 1986 Spatial location of electron trapping defects on silicon by scanning tunneling microscopy *Appl. Phys. Lett.* **48** 724–6

[81] Thmankar R, Raghavan N, Molina J, Puglisi F M, O'Shea S J, Shubhakar K, Larcher L, Pavan P, Padovani A and Pey K L 2016 Single vacancy defect spectroscopy on HfO_2 using random telegraph noise signals from scanning tunneling microscopy *J. Appl. Phys.* **119** 1–8

[82] Grasser T *et al* 2014 On the microscopic structure of hole traps in pMOSFETs *Tech. Dig. Int. Electron Devices Meeting—IEDM14* (New York: IEEE) pp 530–3

[83] Ji J, Wang R, Qiu Y and Huang R 2014 Investigation on random charging/discharging of single oxide trap in SiO_2: an *ab-initio* study *Proc. ICSICT 2014* (New York: IEEE) pp 1419–21

[84] Ji J, Qiu Y, Guo S, Wang R, Ren P, Hao P and Huang R 2014 New framework for the random charging/discharging of oxide traps in HfO_2 gate dielectric: *ab initio* simulation and experimental evidence *Tech. Dig. Int. Electron Devices Meeting—IEDM14* (New York: IEEE) pp 542–5

[85] Uren M J, Kirton M J and Collins S 1988 Anomalous telegraph noise in small-area silicon metal–oxide–semiconductor field-effect transistors *Phys. Rev.* B **37** 8346–50

[86] Grasser T, Waltl M, Goes W, Wimmer Y, El Sayed A-M, Shluger A L and Kaczer B 2015 On the volatility of oxide defects: activation, deactivation, and transformation *Proc. Annual Int. Reliability Physics Symp.—IRPS15* (New York: IEEE) pp 1–3

Chapter 4

Impact device processing and scaling on RTS

RTS is related to the presence of defects in the gate dielectric. The defect concentration depends largely on the processing details (thermal budget, chemistry, stress, processing damage, etc) so the occurrence of RTS will rely heavily on these. At the same time, scaling has some strong implications on the RTS parameters. The relative amplitude is increased for a smaller geometry, while at the same time a decrease of the EOT will increase the screening of the trap potential by the metal gate. As discussed above, RDFs amplify the RTS amplitude, so that all device architectures that implement an undoped silicon channel (FFs, FD SOIs, etc) are beneficial for reducing the impact of single-oxide traps on the transport properties.

4.1 Processing effects on RTS

The effect of the gate oxidation conditions on the appearance of RTS has been noted several times [1–4]. It was shown that more RTSs are found in RTA SiO_2 capacitors in the 600–700°C range (Ar atmosphere) [1]. Nitridation of SiO_2 also results in more RTS [2]. A single oxide (4.8 nm) CMOS apparently has higher and more complex RTS than its dual oxide (0.7 + 4.1 nm) counterparts [4]. RTS in EDT currents in the gate-S/D overlap region have been identified [5–7]. It has also been observed that millisecond junction annealing increases the RTN in the gate current (or bulk oxide trap density) in 1.1 nm SiON transistors [8]. The effect of the silicon substrate orientation on interface states and RTS has been studied by Tega *et al* [9]. They found that the number of RTS traps increases with the amount of suboxides in the interfacial transition layer. The presence of suboxides is largest in Si(110) compared with Si(100) and Si(111) [9]. Moreover, in the case of Si(110) transistors, faster RTSs with larger amplitudes have been observed. This is important for the case of FFs where the main channel is along the (110) sidewalls for a standard ⟨110⟩ channel orientation on a (100) silicon wafer.

It is well-known that separating the channel from the interface using a so-called buried-channel reduces the $1/f$ noise in large transistors and the RTS in small ones

doi:10.1088/978-0-7503-1272-1ch4

[10]. Relatively fast RTSs are suppressed due to the longer tunneling time (larger physical separation). RTS is also present in SOI technologies: partially depleted SOI [11], FD SOI [12, 13] and junctionless (JL) transistors [14, 15]. In the case of FD SOI, it has been demonstrated by experiments and simulation that there is a suppression of extreme RTS events by the low doping density in the silicon film (figure 4.1) [12]. Comparing the 28 and 14 nm FD SOI generations, an improvement has been noted in terms of threshold voltage fluctuations induced by RTS [13]. The same applies for JL transistors, in spite of the very high substrate doping density (of the order $\sim 10^{19}$ to 10^{20} cm^{-3}). In addition, RTS has been studied in silicon-on-nothing (SON) structures, where local SOI can be fabricated by the selective removal of SiGe epitaxial layers and possible refilling with a dielectric [16]. This has been extended to SON nMOSFETs with a HfO_2–SiO_2 gate stack [17]. In recent years, RTS has been studied in extremely thin (ET), highly scaled SOI nFETs [18], showing that in a 14 nm low-power technology it can induce delay uncertainty that exceeds 50% of the nominal delay.

However, in spite of the undoped body, in the case of ultra-scaled planar and tri-gate FD SOI devices, extreme RTS events do occur, as evidenced by the histogram of figure 4.2, obtained after charging the oxide traps with a high gate voltage stress [19]. The exponential distribution and the tail cannot be explained by RDFs or a percolation channel in this case. Also, the MGG cannot account for the observations. According to the simulation results of figure 4.3, the ΔV_T induced by a BOX trap is a factor 2.2 higher on the average than for a gate oxide trap. Similar results have been obtained for a tri-gate transistor. Therefore, the observed distribution tail is ascribed to the effect of BOX traps [19].

Recent studies indicate that the RTS amplitude increases under the influence of compressive mechanical stress in a p-channel transistor, where the stress is induced by nearby STI regions [20]. In the case of 65 nm CMOS transistors with tensile contact etch stop layer (CESL) stress, more RTS noise has been observed compared

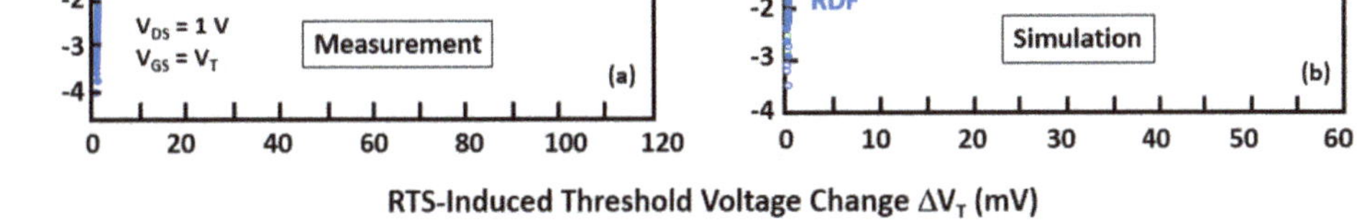

Figure 4.1. Cumulative probabilities of measured (*a*) and simulated (*b*) ΔV_T of bulk and FD SOI MOSFETs. V_{DS} = 1.0 V and V_G = V_T. In the distribution tail, FD SOI MOSFETs have a smaller ΔV_T than bulk MOSFETs. Bulk and FD SOI MOSFETs without assuming RDF are also simulated. (After Nishimura *et al* [12].)

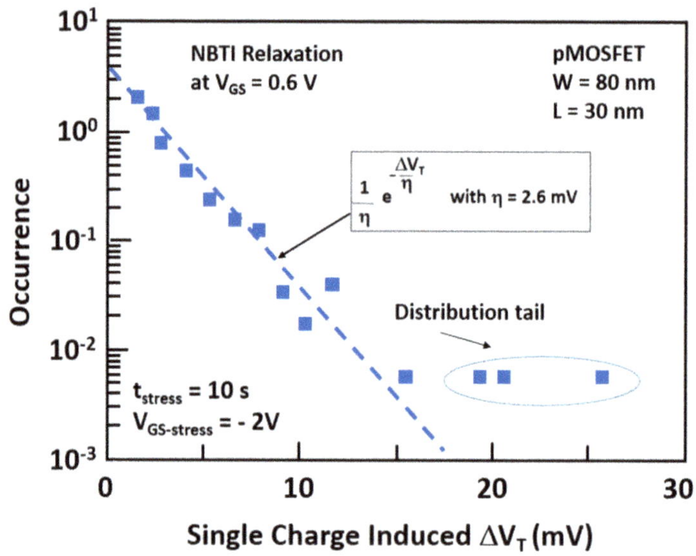

Figure 4.2. Histogram of individual ΔV_T induced by a single trapping event measured during the NBTI relaxation phase at $V_{GS} = -0.6$ V. The ΔV_T distribution follows an exponential law with a mean value η of 2.6 mV. (After Subirats *et al* [19].)

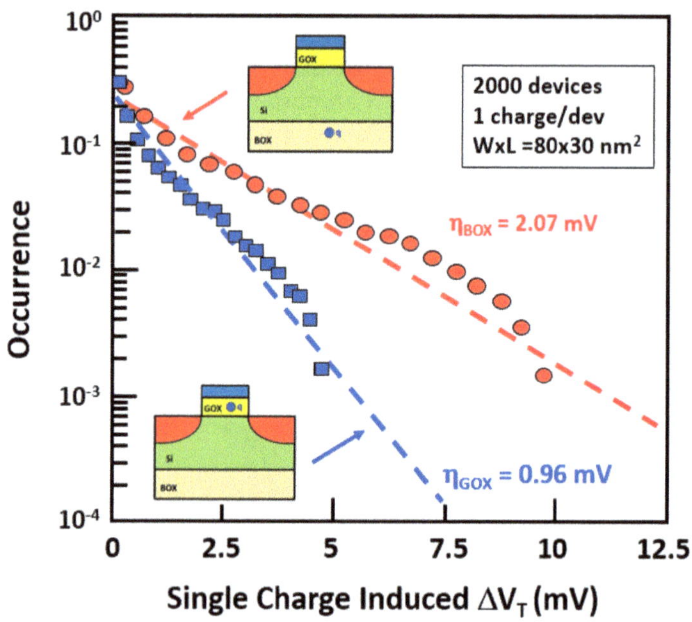

Figure 4.3. Histograms of ΔV_T induced by a charge distributed in the top gate oxide (GOX, squares) or in the BOX (circles). Both can be fitted by a typical exponential law with a 2.2 times higher η for a BOX charge. (After Subirats *et al* [19].)

to devices with SiGe S/D-induced compressive stress [21]. This was attributed to the higher vertical compressive stress in the gate stack. n-channel transistors with compressive (tensile) stress induced by a CESL nitride layer experience larger (smaller) RTN and LFN, respectively [22]. From the time constant ratio, it was concluded that the traps are further away from the interface for tensile stress. On the other hand, a lower $1/f$ (and RTN) noise has been reported for 28 nm pMOSFETs with tip-shaped SiGe S/D [23, 24]. This was assigned to a smaller tunneling attenuation length by the higher compressive stress. Lower $1/f$ noise has also been found in high-κ/metal gate pMOSFETs with embedded S/D stressors [25–27]. At the same time, compressively strained transistors exhibit RTSs which are closer to the Si/SiO_2 interface, compared with the unstrained references. A similar result has been obtained for 28 nm nMOSFETs with tensile strain induced by the stress memorization technique [28].

A large number of reports on RTS in high-κ transistors is now available [6, 7, 29–40]. It has for example been noted that the relative RTS amplitude exhibits a weaker dependence on n_s [32], which was explained by assuming a weaker screening of its charge. This certainly requires further verification. From the strong lattice relaxation in the range of 1.86–2.0 eV, it was derived that the defects responsible for RTS in 1 nm SiO_2/3 nm HfO_2/TaN gate MOSFETs are most likely to be neutral oxygen vacancies (V^0) in the SiO_2 IL [29]. This comes from the fact that HfO_2 scavenges oxygen from the IL, creating oxygen vacancies. This contrasts the behavior of as-grown and stress-induced traps in thick SiO_2 layers, which are characterized by a low relaxation energy (0.13–0.44 eV) and a larger trap depth from the interface [29]. In addition, it has been shown that high-temperature FGA (HTFGA) can decrease the RTS-induced V_T shift by a factor of 2 in 22 nm node compatible high-κ/metal gate transistors [33, 38]. In fact, comparing the cumulative probability distribution in figure 4.4 demonstrates significantly smaller 1σ and 2σ values for the high-κ with HTFGA devices compared to its pure SiO_2 or SiON counterparts [33, 38]. At the same time, a strong geometry dependence of the 1σ and 2σ values has been observed (figure 4.5), which indicates that the RTS-induced

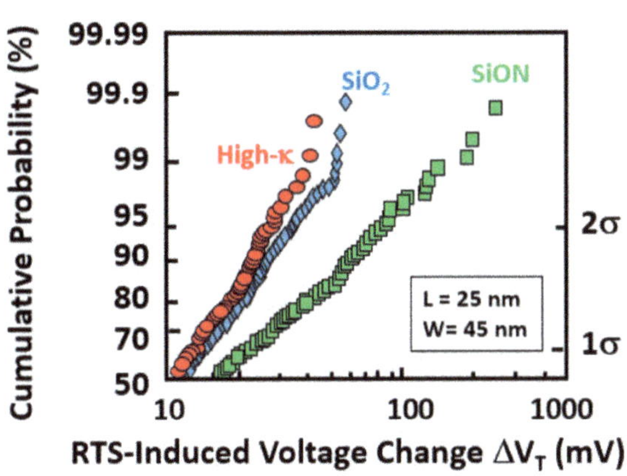

Figure 4.4. Comparison of RTN-induced V_T shift between high-κ with HTFGA, pure SiO_2 and SiON FETs. (After Tega *et al* [33].)

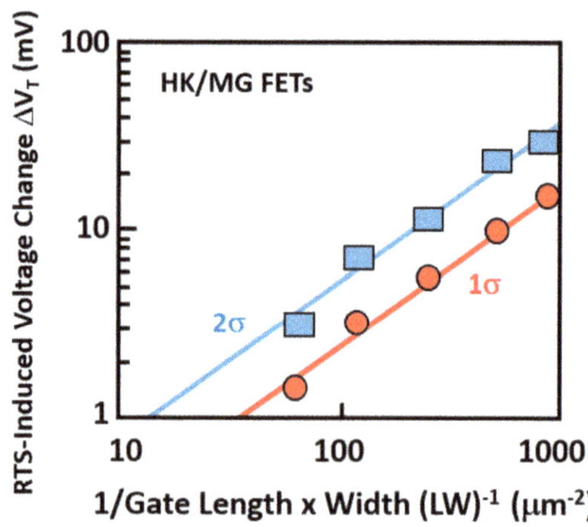

Figure 4.5. Device-size dependence of RTS-induced threshold voltage change ΔV_T. (After Tega *et al* [33].)

V_T shifts will become more pronounced upon scaling and may cross the RDF effect at about the 15 nm technology node [33, 38]. The crossing point has been pushed further in time (from the expected 22 nm node) due to the switch from poly/SiON to a high-κ/metal gate and may even be further shifted by other changes in technology, e.g. replacing the planar architecture by vertical fins.

The effect of post metal-deposition annealing on the RTS behavior of high-κ/metal gate MOSFETs, indicates that passivation can occur through a metal/nitrogen exchange mechanism [41–43]. For the case of mono-metal-annealed (TiN only) devices, traps are located closer to the TiN/HfO_2 interface, due to the substitution of nitrogen atoms by oxygen atoms inside the TiN layer. However, replaced nitrogen atoms from TaN can passivate nitrogen defects in TiN, which improves the device characteristics in dual-metal-layer annealed (TiN/TaN annealed) MOSFETs [41–43].

Some studies have been devoted to the use of Al ion implantation in the high-κ layer of pMOSFETs to tune the effective WF [44–46]. The idea behind this is that Al interacts with oxygen vacancies to form Al–Si–O bonds, thereby inducing a dipole layer at the SiO_2/HfO_2 interface. This changes the effective WF and, hence, the V_T of the pMOSFETs. The gate oxide traps in 28 nm high-κ/metal gate-first technology has been investigated by RTN analysis, showing that the traps are shallower, i.e. closer to the Si/SiO_2 interface in the case of Al ion implantation. At the same time, the barrier for capture E_B was found to be 200 meV higher while the activation energy for re-emission was 200 meV smaller [44]. This has been interpreted in terms of the formation of a thin Al_2O_3 interlayer between the SiO_2 and HfO_2. This results in a higher tunneling barrier for holes, yielding a smaller tunneling depth λ [44].

Comparing RTSs in HfO_2 or $Hf_{1-x}Zr_xO_2$ based nMOSFETs, a different strength of the vibrations of the nearest neighbor Hf atoms to the oxygen vacancy in the immediate vicinity of the oxide trap has been demonstrated [47]. This has been derived from a different gate voltage dependence of the relaxation energy in the multi-phonon emission process. By combining the Arrhenius plot of τ_c and τ_e at different V_{GS}, one can calculate the dependence of the trap binding enthalpy and the activation barrier height for carrier capture E_B, yielding the relaxation energy of the oxygen-vacancy-related traps (equation (2.10)).

In the last five years or so, we have become aware of the fact that RTS can generate functional errors in digital logic circuits and memories (see also chapter 6). There is particular concern about the occurrence of large-amplitude RTSs generating extreme V_T shifts. Such rare events can only be determined by measuring a statistically relevant set of devices (see, e.g. [33, 48–50]), as in a memory. An important conclusion from such a study is that *p-type devices* suffer more from complex RTSs than nMOSFETs [34]. One of the advantages of high-κ gate dielectrics is the possibility for gate WF engineering, allowing a lowering of the doping levels in the substrate, while keeping approximately the same threshold voltage. This should reduce RDFs and also their effect on RTS. The impact of different parameters on the V_T distribution has been experimentally investigated and is reported in figure 4.6 [34]. These distributions can be modeled by a Gumbel distribution, suitable for extreme value statistics. One can also note that the highest V_T shifts are observed for p-channel transistors and for high substrate doping concentration [34, 50]. Therefore, it is possible to state that the dopants act as RTS amplifiers, particularly in pMOSFETs. In fact, an improved channel architecture has been proposed to reduce the effect of RTS, relying on a larger effective channel length and a lower doping density [51].

Further verification of the doping impact has been published by Chen *et al* [52], showing at the same time that a combination with strain engineering suppresses the RTS amplitude. This is related to the fact that, in general, the $\Delta\mu/\mu$ contribution to

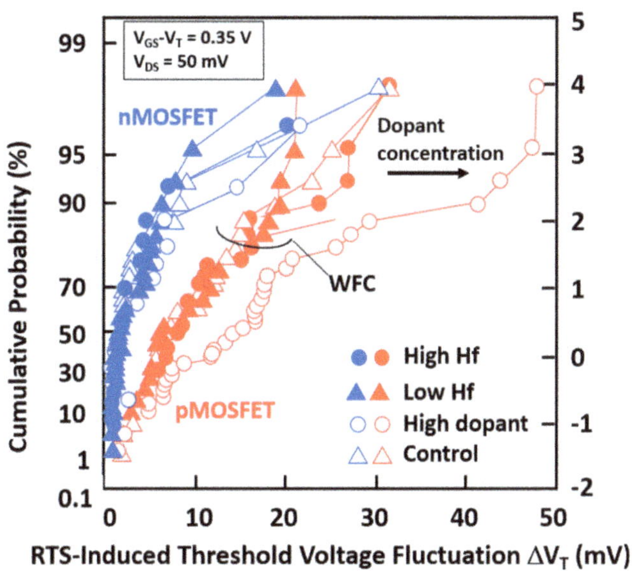

Figure 4.6. Cumulative distributions of V_T fluctuation due to RTS. $V_{GS} - V_T = 0.35$ V, $V_{DS} = 0.05$ V and sampling rate 10^5 samples/s. The devices have a polysilicon gate, while the controls have SiON (1.85 nm EOT). The role of Hf is to control the gate WF by utilizing Fermi-level pinning at the gate/dielectric interface. (After Shimizu *et al* [34].)

$\Delta I_D/I_D$ is dominant and is reduced for a higher low-field mobility by strain engineering.

A deeper analysis of the data underlying figure 4.6 demonstrates that the scattering coefficient is not affected by the dopants or the WF engineering (Hf content), as clearly illustrated in figure 4.7, where the different curves are nearly falling together. The trap number term, on the other hand, represented by the flat-band voltage fluctuation ($\sim qN_{ot}/C_{ox}$) in figure 4.8 is very sensitive to the process conditions. This leads to the conclusion that the longer tail distributions in V_T are due to a higher number of traps in such an extreme value transistor. This could for example come from dopant implantation, which increases N_{ot} with a higher dose. WF control appears to reduce the number of (active) traps [34].

Another statistical analysis of the effect of RTS on a large set of CMOS transistors has revealed a clear asymmetry along the S/D direction [53]. This asymmetry is caused by the in-plane non-uniform distribution of traps and becomes more pronounced for shorter transistors. According to figure 4.9, a SiON gate dielectric has a more pronounced tail of extreme V_T values compared with SiO_2. This indicates a higher number of complex or anomalous RTSs per transistor or more transistors with more than one active trap. On the other hand, plasma damage did not change the threshold voltage distribution, based on the results obtained on structures with different antenna ratios [53].

In addition to the device geometry and architecture, the gate dielectric (and gate stack) and the substrate doping density, another factor which clearly plays a distinct role is the device isolation at the edges of the transistor. It has been demonstrated that either in LOCOS [54] or STI isolated transistors [55], RTS can preferentially occur at the transistor edges. This can be related to dopant pile up (and fluctuation),

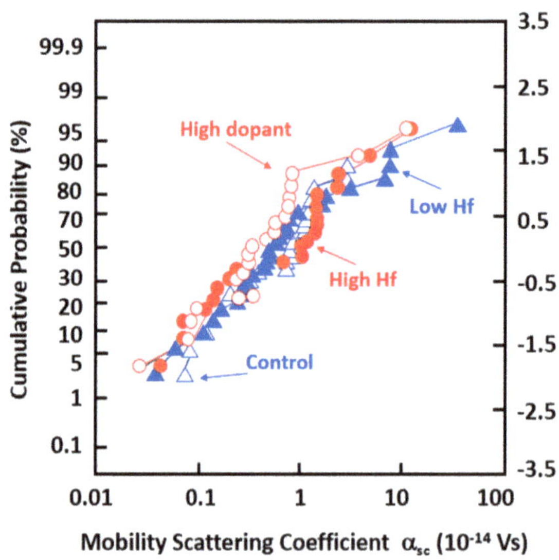

Figure 4.7. Cumulative distributions of mobility scattering coefficient in pMOSFETs. (After Shimizu *et al* [34].)

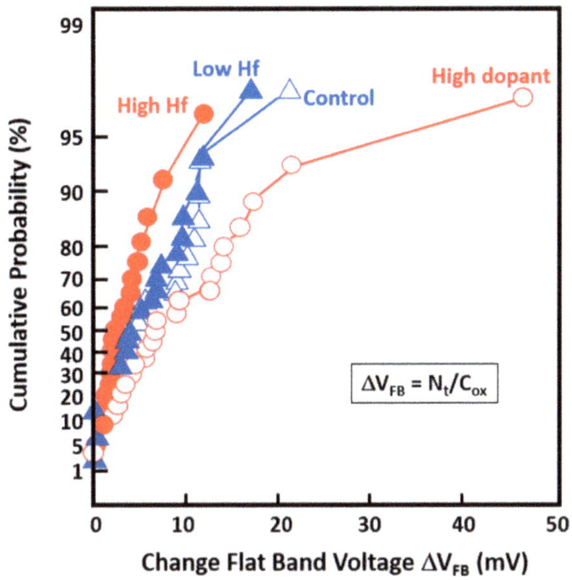

Figure 4.8. Cumulative distributions of flat-band voltage fluctuations in pMOSFETs. (After Shimizu *et al* [34].)

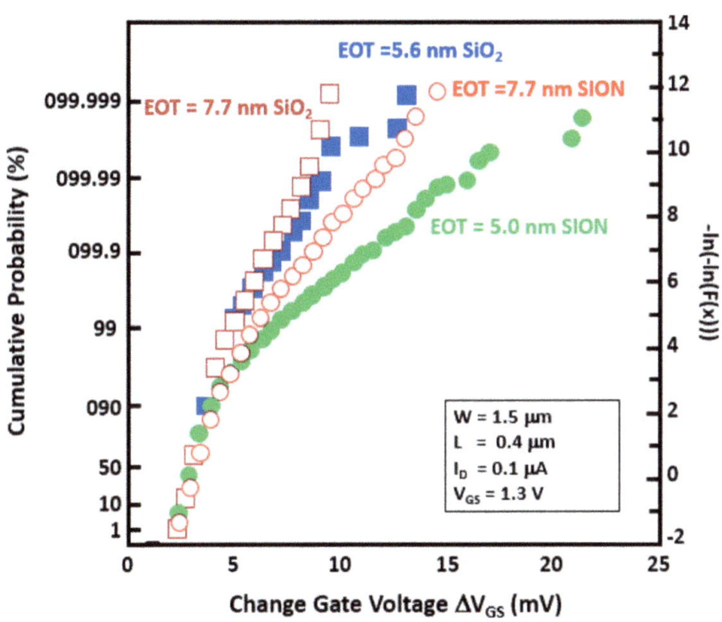

Figure 4.9. Cumulative distributions of ΔV_{GS} represented by a Gumbel plot for different gate insulators, which include SiO_2 with 5.6 and 7.7 nm EOT and SiON with 3.0 and 7.7 nm EOT, respectively. These samples were measured with appropriate biases to equal the carrier densities among each sample. (After Abe *et al* [53].)

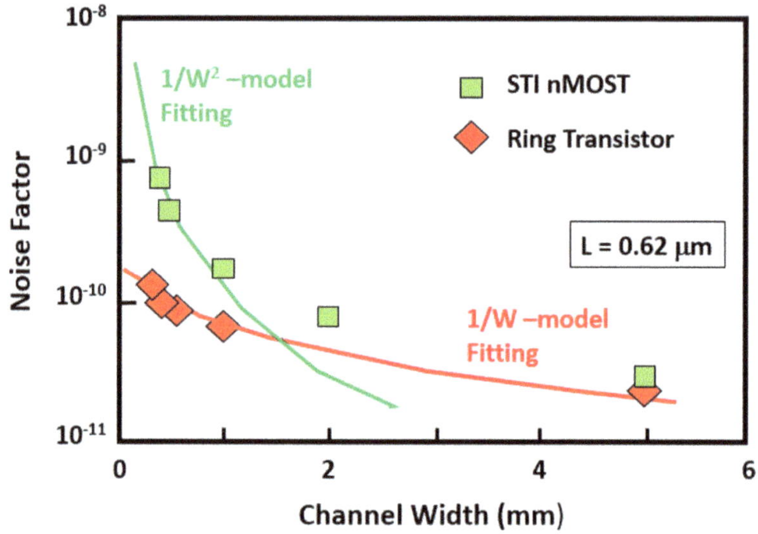

Figure 4.10. Noise factor as a function of channel width (channel drawn length = 0.62 μm) when biasing in the linear regime for both ring and STI-nMOS transistors. the noise factor is the average of the integrated noise power spectra (S_{ID}/I_D^2) × f across the whole frequency range. The noise factor is higher in STI-nMOS transistors, in particular for narrow-width transistors. Also shown in the graph are model fitting lines with $1/W$ and $1/W^2$ relationships. (Reproduced with permission from Wang *et al* [55]. Copyright 2009 IEEE.)

leading to PPs along the device edges or process damage (stress) associated with the poorer quality thick isolation oxide. In the case of STI this could result from the dry etching and polishing (chemical mechanical polishing) steps [55]. Figure 4.10 nicely illustrates that the RTS noise factor in a ring device without STI is much lower than for a device with STI [54].

4.2 RTS in fin-type architectures

In state-of-the-art and the next CMOS technology nodes based on SOI or bulk FFs, conduction occurs along a different (110) crystallographic plane compared to the standard (100) plane. A recent detailed comparison of RTS in (100) and (110) silicon has revealed that neutral traps are dominant in (110) devices compared to the standard orientation [56]. The type of trap (neutral versus attractive) was distinguished based on the Coulomb blockade effect and the energy term ΔE, derived from the capture and emission time against the drain current [57]. Neutral traps evolve proportional to n_s^2, while attractive traps go with n_s. It was also shown that neutral traps produce a higher ΔV_T, so that their impact is more pronounced (figure 4.11) [56]. There are also more neutral traps in p- than in n-channel transistors for the technology studied. According to figure 4.11, the hole traps are further from the interface. Attractive traps have a 60–100 meV lower activation energy (closer to E_F) than neutral centers for both hole and electron traps. From these results it has been concluded that RTS-induced V_T shifts are much more pronounced in standard orientation pFFs (figure 4.12) [27], while these can be suppressed by a proper choice of channel orientation for nFFs.

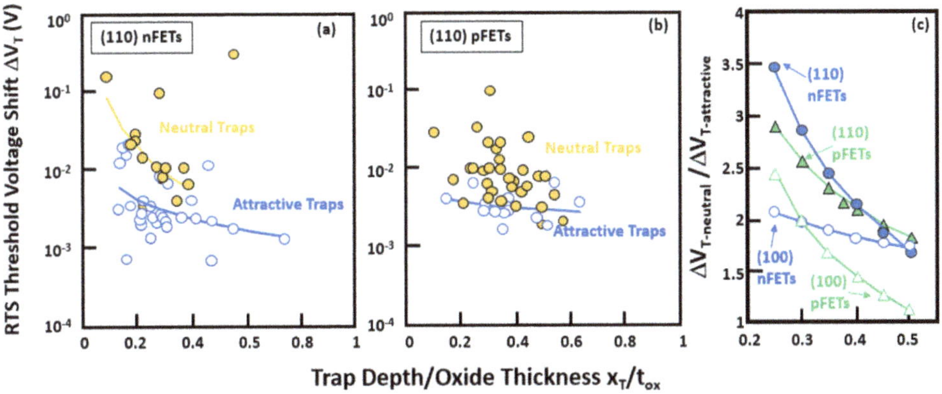

Figure 4.11. Scatter plot of trap positions and threshold voltage shifts ΔV_T caused by RTS in (*a*) nFETs and (*b*) pFETs. Neutral traps show larger ΔV_T than attractive traps do. (*c*) Calculated ΔV_T ratios of neutral traps to attractive traps, using power-law approximation fittings in (*a*) and (*b*). (Reproduced with permission from Chen *et al* [56]. Copyright 2012 IEEE.)

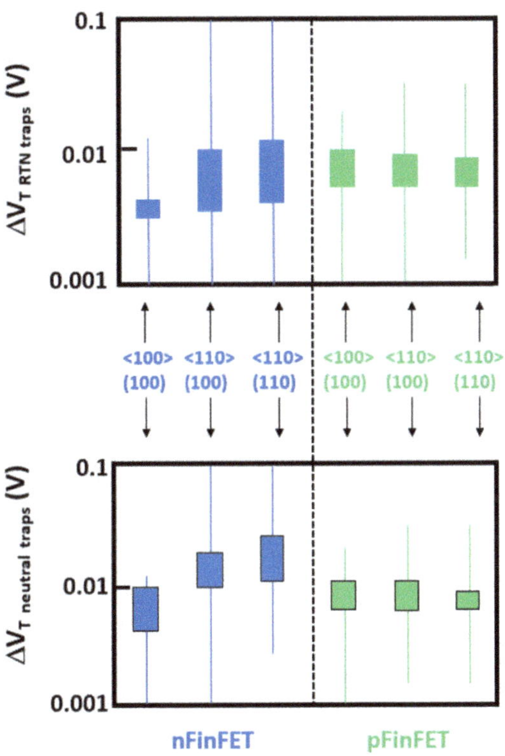

Figure 4.12. Estimated ΔV_T values in n/p-FFs from all RTS traps (top), and just from neutral traps (bottom). (Reproduced with permission from Chen *et al* [56]. Copyright 2012 IEEE.)

A further numerical modeling of the effect of the trap location on the RTS and LFN of high-κ/metal gate FFs has been reported by Lee *et al* [58], demonstrating the effect on the noise variability of localized traps.

4.3 Nanometric scaling aspects of RTS

The current 14 nm CMOS technology node is based on a (bulk) silicon FF architecture with a high-κ/metal gate and is expected to be further scaled down to at least the 7 nm node. In parallel the FD SOI planar counterpart is keeping up with the same scaling trend and, although somewhat less popular, it may become the dominant technology for Internet of Things applications. The success of the FF lies in the excellent gate control over the channel electrostatics, thereby minimizing short-channel effects. The ultimate performance enhancement is expected for a so-called gate-all-around (GAA) architecture around a planar or vertical NW, for which the channel material is still a matter of debate: either silicon or so-called high-mobility semiconductors ((Si)Ge for pFETs and InGaAs, InAs, etc, for nFETs). In parallel, tunnelFETs (TFETs) are also currently the subject of intense study to replace the traditional drift-diffusion MOSFET. The operation principle of a TFET, based on band-to-band tunneling of carriers through a (gated) p–i–n structure, enables one to obtain an SS below the thermal limit of 60 mV/dec at room temperature. This allows a decrease of the supply voltage and is favorable for the low-power operation of future deeply scaled circuits. In this context one can ask the question whether the RTS behavior will be affected by these different architectures, channel materials and operation principles.

4.3.1 Scaling trend RTS amplitude

Based on the extensive work carried out in the past, it should be possible to theoretically foresee the impact of RTS in future nanoscale transistors. As shown in figure 4.13 a very high relative amplitude RTS can be reached in a 35 nm × 85 nm transistor [59], which can be approximated in a simple way by the 'hole-in-the-inversion-layer' model. It is clear that the highest amplitude is found for low gate voltages (no free carrier screening), while at the same time, EOT scaling results in a reduction of the worst-case amplitudes (figure 4.14).

The drain current for a trap with dimensions $L_t \times W_t$ is given by [60]

$$\frac{\Delta I_D}{I_D} = \frac{L_t W_t(\sigma - \sigma_t)}{WL(\sigma - \sigma_t) - W_t(L - L_t)(\sigma - \sigma_t)} \tag{4.1}$$

with σ and σ_t the channel conductivity outside and inside the trap area, respectively. In figure 4.13 it has been assumed that $L_t = W_t$ and that the channel conductivity inside the squared trap is zero. For that case the half hole size, i.e. the radius of a circle within a squared trap, is calculated and shown in the figure.

The same approach has been extended to evaluate the impact on so-called high-mobility channel devices [60]. The key result of this study is summarized in figure 4.15. It is shown that for silicon channels the worst-case RTS amplitude is beyond 10% at the 16 nm node. At the same time, the RTS amplitude becomes

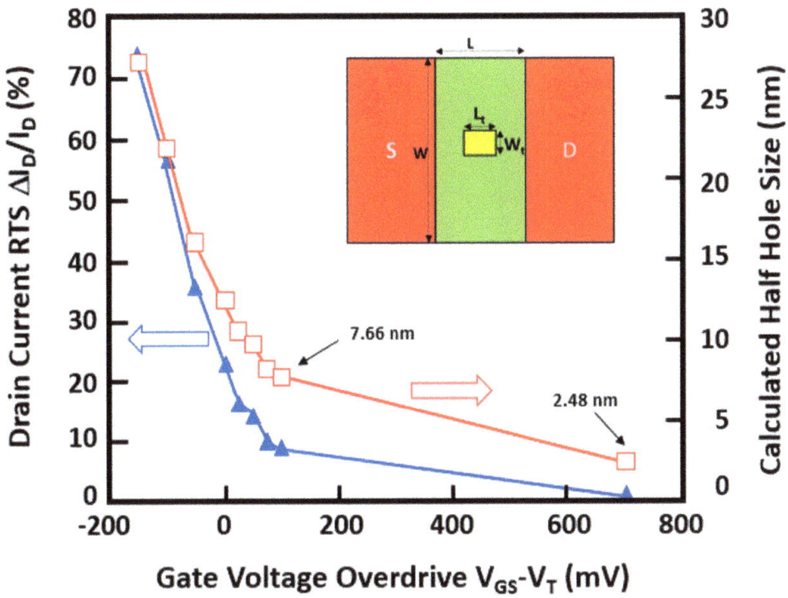

Figure 4.13. The measured RTS amplitude (triangles) is a smooth function of the gate voltage overdrive. The inset shows a trap with dimensions $L_t \times W_t$, called the cored-out section. The calculated half cored-hole size, i.e. the radius of a circle within a squared trap, drops rapidly as the gate voltage overdrive increases from below to above threshold. The as-drawn size of the device is 55 nm × 85 nm, while the size used for the calculation is 35 nm × 85 nm. (After Cheung *et al* [59].)

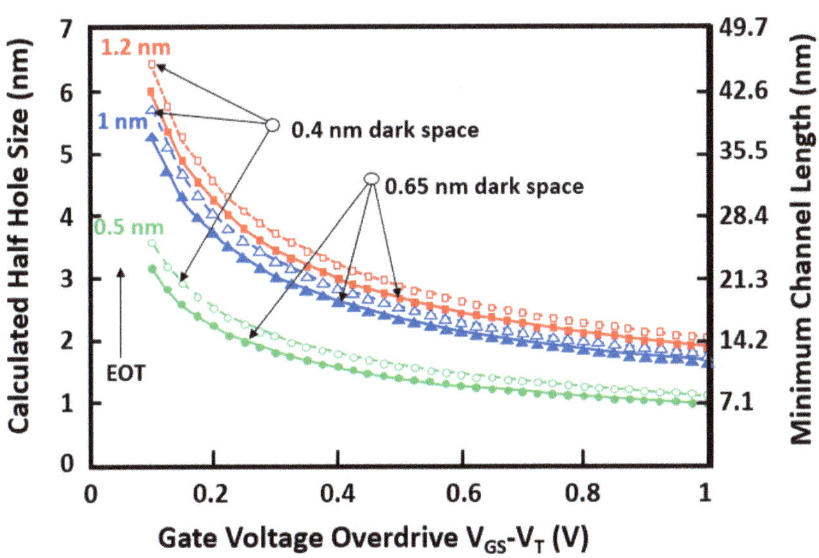

Figure 4.14. Half cored-out hole size as a function of gate voltage overdrive for three different oxide thicknesses and for 0.4 nm dark space (broken line) as well as for 0.65 nm dark space (solid line). (After Cheung *et al* [59].)

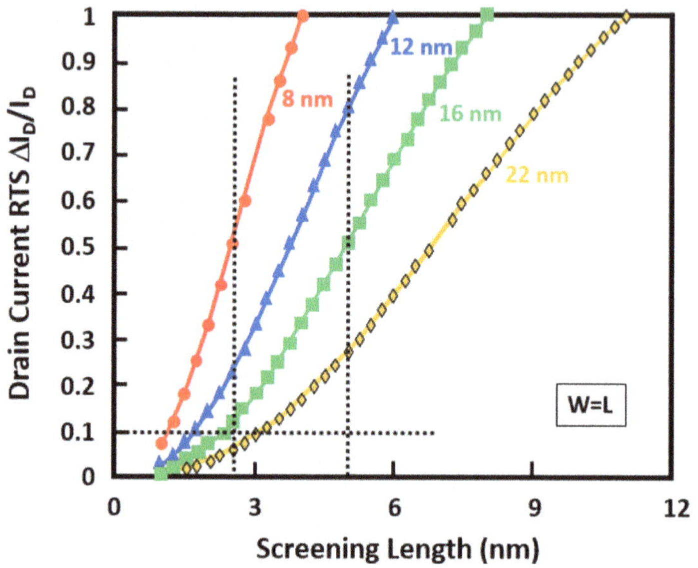

Figure 4.15. Worst-case RTS amplitude as a function of screening length for various technology nodes. The first broken vertical line marks the experimentally determined screening length for silicon. The second broken vertical line doubles the screening length as a result of reducing the inversion charge density by a factor of 4. (After Cheung *et al* [60].)

extreme when the screening length increases due to the decrease in inversion charge density, as would be the case for high-mobility channel structures. As a result, the size of the cored-out area is determined by the screening length, according to equation (4.2). This is indicated by the vertical lines in figure 4.15.

$$\frac{\Delta I_D}{I_D} = \frac{p'^2}{n\left(WL - \frac{L}{p'\sqrt{n}} + p'^2 n\right)} \tag{4.2}$$

with n the inversion carrier density $\cong C_{ox,eff}(V_{GS} - V_T)$ and the screening length $W_t = L_t = p'/\sqrt{n}$; p' being the proportionality constant. At low drain currents, the presence of a PP yields a narrow effective width and an artificially large screening length. As the gate voltage overdrive increases, the PP widens and eventually disappears. The screening length behavior of figure 4.13 is largely due to a widening of the PP under the cored-out hole. This concept can be approximated by the hole in the channel with size $W_t \times L_t$ and reduced (or zero) conductivity, whereby the screening length determines the size of the hole.

Different geometrical laws have been predicted for the geometry scaling of the average ΔV_T induced by a single RTS. Usually, it is concluded that ΔV_T is inversely proportional to $W_{eff}L_{eff}^{0.5}$ [49]. However, for FD SOI, an inverse dependence on $W_{eff}L_{eff}$ has been reported [19]. Additionally, numerical investigations have shown that the area dependence of the RTS amplitude depends on the operation bias and changes from $(W_{eff}L_{eff})^{-0.5}$ in the subthreshold regime to $1/(W_{eff}L_{eff})$ at high currents [61].

The area scaling trend of the RTS amplitude has also been investigated for the case of FFs using atomistic device simulations [62]. Contrary to what is expected from planar devices, an inverse area scaling law is obtained for FFs, whereby the RTS amplitude decreases for a smaller device area. This is explained by considering the position of the inversion layer charge. In weak inversion (nFET at low V_{GS}) where the RTS amplitude is normally the highest, the inversion charge is concentrated in the center of the fin away from the interfaces, which reduces the impact of an oxide trap on the channel current [62]. At higher V_{GS} (high I_D), the centroid moves closer toward the interfaces, which increases the impact of the trap charge but, at the same time, Coulombic screening by the mobile carriers reduces the effect. In planar MOSFETs, the charge centroid is always close to the interface and is barely modulated by the gate voltage.

4.3.2 Silicon GAA NWs

Further along the International Technology Roadmap for Semiconductors (ITRS), NWFETs may offer a solution for scaling to the 10 nm and beyond CMOS nodes. RTS has been studied in NW transistors [63–74], exhibiting huge relative amplitudes. More recently, the importance of Coulomb blockade has also been highlighted for RTS occurring in GAA silicon NW transistors (SNWTs) [68]. In that case, the device may have a circular geometry with a radius R [66, 70]. Appropriate equations for the time constant ratio have been worked out, leading in this case to [66]

$$x_T = \left\{ \left(1 + \frac{t_{ox}}{R}\right)^{\frac{\left(\frac{kT}{q}\frac{\partial ln\left(\frac{\tau_c}{\tau_e}\right)}{\partial V_{GS}} + \frac{\partial \aleph_S}{\partial V_{GS}}\right)}{\left(\frac{\partial \aleph_S}{\partial V_{GS}} - 1\right)}} - 1 \right\} R \tag{4.3}$$

The equivalent of equation (2.15) for the lateral trap position y_T, based on measurements of the time constant ratio as a function of the drain voltage in forward and reverse operation, has also been worked out along the same lines [70]. Similar equations have been derived for a recessed channel MOSFET with a circular geometry [67].

In addition, it has been pointed out that quantum confinement will affect the RTS parameters, in particular the time constants (figure 4.16) [71]. It turns out that the bias dependence becomes more pronounced when the temperature dependence is weaker in figure 4.16. Another effect which starts to play a role is the non-stationarity of the RTS at high operating voltages (high electric fields). Under such non-stationary conditions, the capture time constant is no longer exponentially distributed but rather follows a Weibull distribution (figure 4.17(*a*)) given by [71]

$$P = 1 - \exp\left[-(t/\tau_c)^{\beta}\right] \tag{4.4}$$

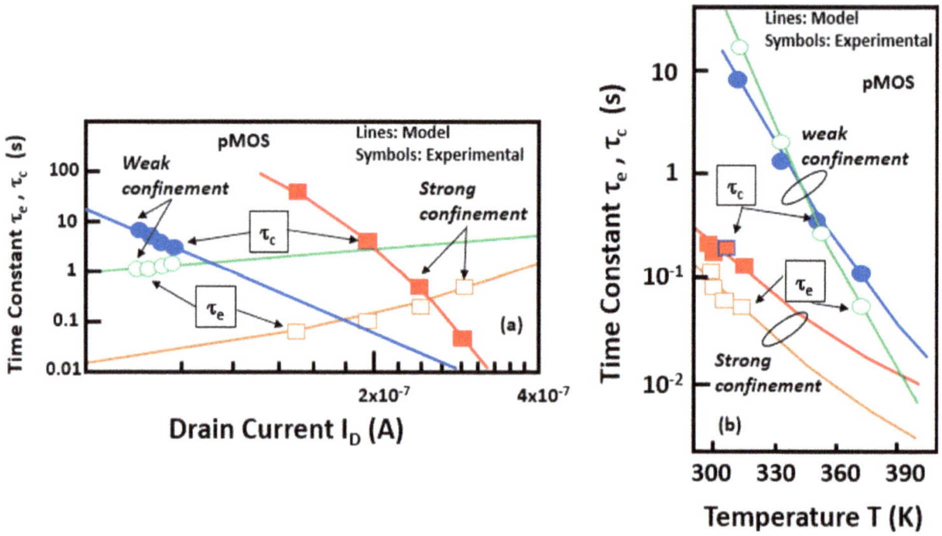

Figure 4.16. (*a*) Typical experimental results (symbols) of the measured capture time constant $\langle \tau_c \rangle$ and emission time constant $\langle \tau_e \rangle$ with varying drain current (i.e. gate bias) of RTS in a SNWT with large diameter (~80 nm for weak confinement) and small diameter (~10 nm for strong confinement). The lines are the modeled parameters, including quantum confinement. (*b*) Typical experimental results (symbols) of the measured $\langle \tau_c \rangle$ and $\langle \tau_e \rangle$ with varying temperature of RTS in SNWTs with large and small diameters. (After Liu *et al* [71].)

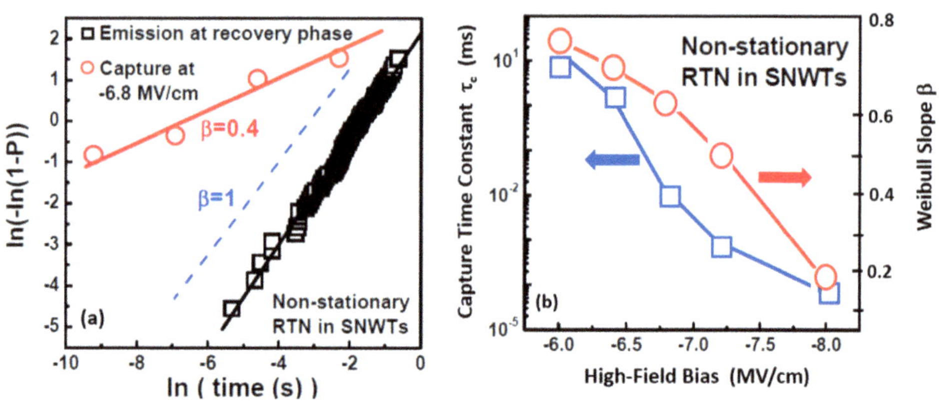

Figure 4.17. (*a*) The extracted capture and emission probability of non-stationary RTS in SNWTs biased at -6.8 MV cm^{-1} as a function of time. The emission time of non-stationary RTS still follows the exponential distribution, while the capture probability obeys the Weibull distribution. (*b*) The capture time constant of the non-stationary RTS in SNWTs as a function of high-field bias, extracted from the measured capture probability distribution in (*a*) and the extracted Weibull slope β as a function of high-field bias. The $\langle \tau_c \rangle$ of non-stationary RTS decreases with increasing high-field bias. (After Liu *et al* [71].)

Equation (4.4) reduces to an exponential distribution for $\beta = 1$, while for a Weibull distribution $\beta < 1$. A dedicated, so-called statistical trap-response method has been developed for investigating non-stationary RTS parameters [71]. As shown in figure 4.17(*b*) both τ_c and β are shown to decrease with increasing electric field.

A detailed study of the geometry dependence of the LFN in tri-gate NWFETs has revealed that pFETs are less susceptible to RTS than nFETs [72]. In summary, the width dependence of the RTS amplitude in tri-gate NW MOSFETs follows a $1/W^{0.5}$ dependence, with W equal to the fin width plus two times the height and a $1/L^{0.5}$ length dependence [75]. This yields a similar area dependence as for a planar transistor. At the same time, no effect of the surface orientation of the NW, i.e. ⟨100⟩ versus ⟨110⟩ was observed [75]. Based on the experimental observations, a strategy has been proposed to reduce the noise/RTS in such devices, emphasizing the role played by dopants in the channel. It has also been shown that using a four-probe NW transistor, one can study the channel potential fluctuations induced by an RTS [76]. It was found that RTSs cause potential fluctuations in the entire channel region.

With respect to RTS as a source of variability in an SNWT, it has been shown that source-induced RDFs overwhelm RTS and become even more dominant for shorter lengths [77]. Other studies revealed opposite trends with respect to the diameter dependence of the noise, where an increase with d_{NW} is found for nFETs and a decrease for pFETs [78]. It was also found that switching gate bias operation reduces the noise magnitude in nFETs dramatically.

Due to the high surface-to-volume ratio, silicon NWFETs are extremely sensitive to a single charge in the gate oxide, which can be exploited in the development of biosensors [63, 64, 79, 80]. For example, single-trap phenomena in liquid-gated Si NWFETs can be used as an effective and versatile approach for this purpose. It turns out that the switching kinetics of current fluctuations generated by a single trap in the dielectric are very sensitive to the pH in the liquid gate [79]. This is related to the change in the carrier density induced by the difference in the WF potential of the gate.

For ultimately scaled NWFETs, where the device size becomes smaller than the Debye screening length, the impact of a single oxide charge on the device current becomes [80]

$$\frac{\Delta I}{g_m} = \frac{q^*}{C_G} \tag{4.5}$$

with q^* the effective charge on the trap, considering Coulomb repulsion and C_G the gate capacitance. It turns out that the usual number fluctuation model (equations (2.26) and (2.27)) will overestimate the effect of the RTS (i.e. $q^* < q$) [80]. The upper limit for the noise in nanoscale devices can be derived by considering the noise from a single trap with the trap occupation probability equal to 50%, corresponding to $S_{VG} = 0.08q^{*2}/C_G^2 f$. The RTS amplitude and the normalized Coulomb repulsion potential are linked together, which allows an analytical derivation for q^*, namely,

$$\frac{q^*}{q} = \frac{2\varepsilon_{ox}}{\varepsilon_{ox} + \varepsilon_S}\left[\frac{1}{1 + \frac{qt_{inv}}{2kT\varepsilon_0\varepsilon_{ox} + \varepsilon_S}C_{ox}\left(V_{GS} - V_T\right)}\right] \tag{4.6}$$

This corresponds to the elementary charge trapped in the oxide minus its image charge in the semiconductor [80]. Integration of equation (4.6) over a density of traps N_{ot} (cm^{-2}) homogeneously spread in position and energy gives [80]

$$S_{VG} = \left(\frac{2\varepsilon_{ox}}{\varepsilon_{ox} + \varepsilon_S}\right)\frac{q^2 N_{ot}}{C_{ox}^2 WL}\left(\frac{1}{1 + \frac{qt_{inv}}{4kT\varepsilon_0\varepsilon_S}C_{ox}\left(V_{GS} - V_T\right)}\right)\frac{1}{f} \tag{4.7}$$

Equation (4.7) is an alternative form for the so-called correlated-mobility number fluctuation $1/f$ noise model, which is more physics-based.

RTS have also been studied in arsenic-doped polysilicon NWFETs at low temperatures [81]. Under a strong magnetic field, the states of the As impurities split up, coexisting with the two-level state of an oxide trap.

4.3.3 High-mobility channel materials

There is currently a strong interest in the development of high-mobility channel transistors to further boost device performance. The impact of advanced technologies on the number and average amplitude of RTS in nanoscale transistors has been studied experimentally and using 3D simulations in great detail by Franco *et al* [82–85]. It has been observed that in high-κ/metal gate bulk Si pFF structures (capacitance equivalent thickness (CET) ~ 1.7 nm) the geometry dependence of the average RTS voltage amplitude η can be expressed as 1/Area, which is more severe than for a planar flash memory transistor. This illustrates the better electrostatic control of the gate over the channel area in the tri-gate architecture [82]. At the same time, the average voltage amplitude is two times higher than expected for a single charge value q/C_{ox} [82]. In addition, SiGe channel planar pFETs show a strongly decreased (<10%) average number of active defects $\langle n_t \rangle$ (figure 4.18) in combination with a decreased η, even when normalizing for the reduced CET. This suggests fewer active traps in the buried SiGe channel [82, 84]. One can read from figure 4.19 that the average oxide-trap density decreases with thinner Si cap layer thickness [82, 84]. This can be explained by the favorable alignment of the Fermi level in the SiGe quantum well with respect to the band of trap states in the gate dielectric (figure 4.19) [84].

Studying the gate voltage overdrive dependence of the ΔV_T impact for a large number of RTSs in similar nanoscale pMOSFETs, one can observe in figure 4.20 a different behavior for all the different traps. From 3D atomistic simulations it follows that a maximum ΔV_T at threshold voltage indicates a trap right above the PP in the channel, while if the maximum RTS amplitude is away from V_T, the trap is away from the critical spot (figure 4.21). 3D atomistic simulations indicate that the highest RTN-induced threshold voltage shift is strongly reduced in real operating conditions [82].

Recently, results on RTS in Ge [86] and near-ballistic InGaAs GAA NWFETs [87–89] have been reported and many are expected to follow.

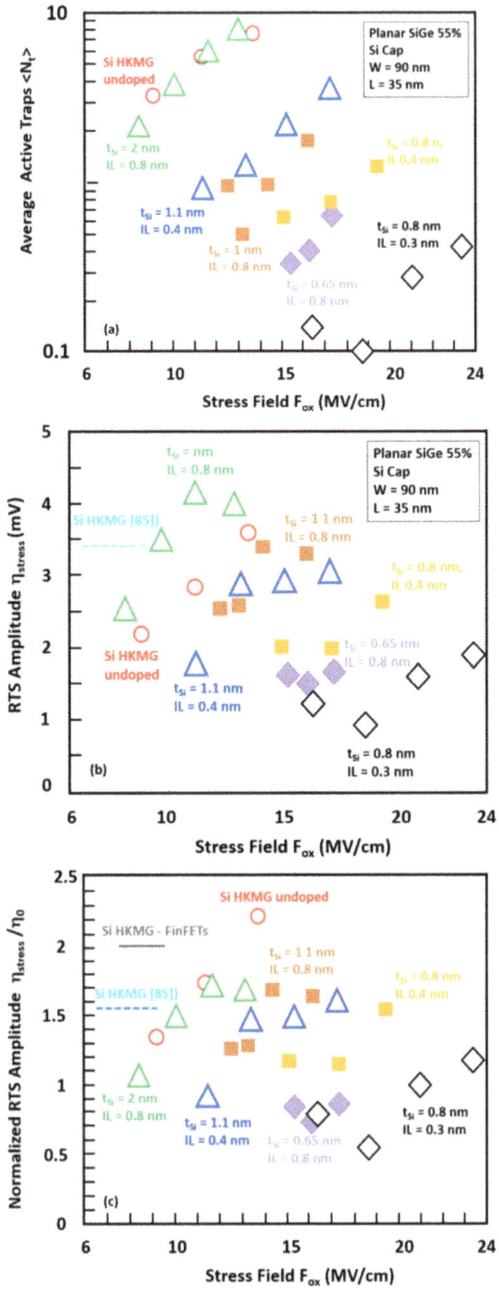

Figure 4.18. Technology comparison from the perspective of the impact of individual trapped charge. (*a*) The observed average number of active defects ($\langle n_t \rangle$) is strongly decreased for optimized undoped SiGe channel pFETs with a reduced Si cap thickness, even with respect to an undoped silicon reference. (*b*) The same SiGe devices also show reduced η_{stress} values. For a fair comparison, (*c*) the η_{stress} values are normalized w.r.t. η_0 (unstressed RTS voltage amplitude) to account for the different C_{ox} between devices of different technologies. SiGe devices still show the best η/η_0 values. (Reproduced with permission from Franco *et al* [82]. Copyright 2012 IEEE.)

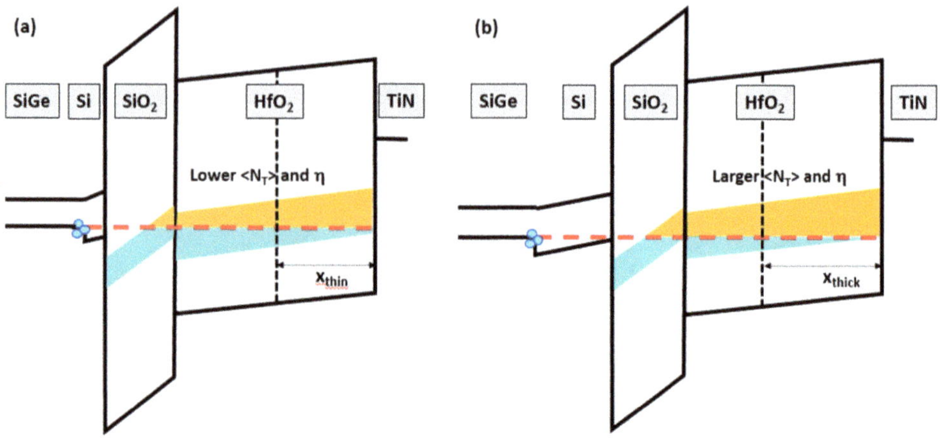

Figure 4.19. Sketches of the SiGe gate stacks in the stress condition, showing the defect bands in SiO_2 and HfO_2. Due to the higher Fermi level in the SiGe channel, fewer defects are energetically favorable for channel holes. Moreover, the accessible defects are preferentially located further from the channel, toward the gate, yielding a reduction of η. (*a*) Favorable alignment shift is maximized for the thinnest Si cap, while (*b*) the additional voltage drop over a thicker Si cap (at fixed field) makes more defects on the channel side accessible. This readily explains the larger $\langle n_t \rangle$ and η values observed for SiGe devices with thick SiGe caps. (After Franco *et al* [84].)

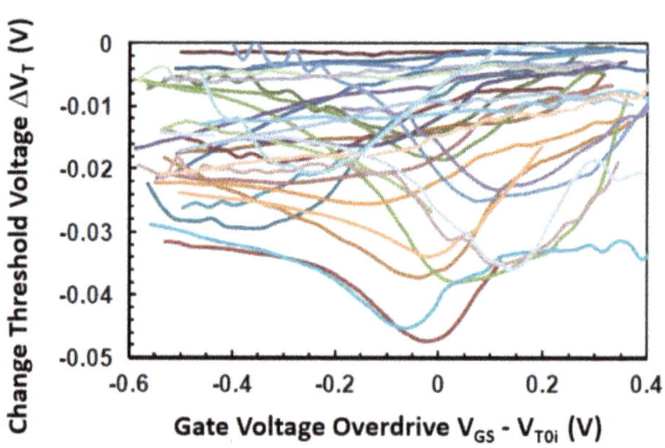

Figure 4.20. Different $\Delta V_T(V_G)$ defect characteristics are found when measuring 30 different (but nominally identical with V_{T0i}) devices before and after the capture of a single hole. Note that these devices were chosen to show large impact of the single charge gate oxide defect and do not represent the entire population. (Reproduced with permission from Franco *et al* [82]. Copyright 2012 IEEE.)

4.3.4 RTS in TFETs

TFETs on SOI are also susceptible to RTS with a large amplitude, which can be explained by the small effective gate oxide area [90]. A numerical analysis has demonstrated that huge RTS amplitudes of several hundreds of per cent can be achieved, due to the enhancement of the band-to-band tunneling current by the impact of a charged trap on the tunneling probability, by shortening the critical tunneling path [91]. The critical factor is the junction electric field where electron–hole

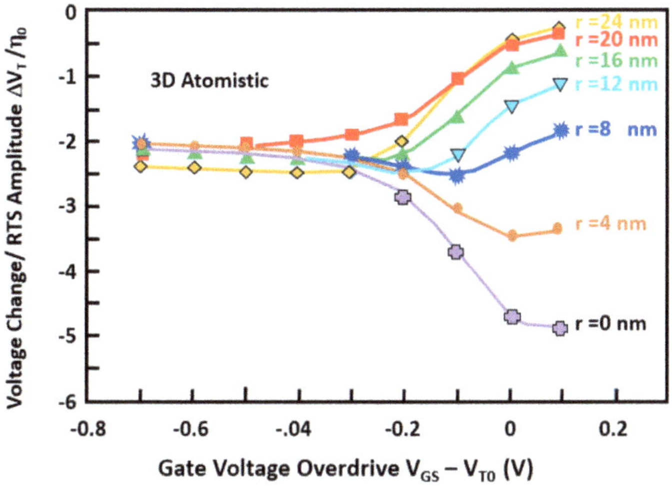

Figure 4.21. 3D atomistic simulation results: different $\Delta V_T(V_G)$ characteristics are expected as a function of the distance with respect to the critical PP: a maximum ΔV_T for $V_G \approx V_{T0}$ is ascribed to a charged trap located directly above the critical spot, while a larger ΔV_T for increasing absolute gate voltage is expected for a trap located away from the critical spot. (Reproduced with permission from Franco *et al* [82]. Copyright 2012 IEEE.)

pairs are generated by non-local band-to-band tunneling and not the maximum electric field [92]. A large variability in noise behavior has been observed, with little dependence on the gate length. The noise variability can be improved by a better control of the junction abruptness [92].

4.4 RTS in 'beyond-silicon' devices

While there is a reasonable consensus about the upcoming CMOS technology nodes up to 7 nm, with silicon bulk FFs the frontrunner for the moment and (Si)Ge and possibly III–V channels may be used at some point in a GAA NW configuration, further on along the ITRS, the possibilities are still open. In addition to TFETs, alternative materials, such as CNTs, graphene and other 2D semiconductors (MoS_2, $MoSe_2$, WS_2, etc) are in competition. Even further ahead, several options for single-electron and quantum (spin) computing are being intensely investigated. It is obvious that, given the small size of the active devices, traps and RTN will be even more important than for today's technologies. Here, the available data on exploratory devices and architectures will be summarized.

4.4.1 CNT FETs

Initial results showed that in CNT FETs, giant RTSs can be observed [93–97]. The huge amplitudes typically found, have been interpreted in terms of a higher scattering coefficient (10^{-12} to 10^{-13} Vs) [93], indicating a strong coupling of the trapped charges in the gate oxide wrapped around the nanotube to the channel. A distinction has to be made between RTS in semiconducting and metallic CNTs and whether long-channel (diffusive transport) or short-channel (ballistic transport)

CNT FETs are considered. Usually, semiconducting CNT FETs are p-type. In a diffusive semiconducting CNT FET, the main effect of RTS is associated with the mobility and not so much with the carrier number or V_T shift [95]. Also a different V_{GS} and V_{DS} dependence of the RTS amplitude with respect to conventional MOSFETs has been reported [95]. This is, among others, related to the fact that it is difficult to fabricate good ohmic contacts to CNTs. Metallic contacts tend to have some barrier which impacts on the alignment between the trap level in the oxide and the Fermi level in the CNT channel [96] (figure 4.22). It should be noted that the band gap of a semiconducting CNT scales with $1/d_{CNT}$, with d_{CNT} its diameter. This also gives rise to a different bias dependence of traps close to the contacts or in the middle of the channel [95].

From the experimental emission and capture time constants, shown in figure 4.23 [96], one can derive the trap depth from the interface in a similar way as for a MOSFET, e.g. using equation (2.13), provided $\partial E_F/\partial V_{GS}$ is negligible (in strong inversion) [96]. For RTS measurements performed at low T (4.2 K) one should also account for the fact that the electron temperature may be different than the lattice temperature, at large S/D biases. The Coulomb potential of a charged trap near the interface can (nearly) completely block off the channel current, in particular for small diameter CNTs. For diffusive CNT FETs, the RTS amplitude is due to mobility modulation by Coulomb scattering [96]. For short CNTs, on the other hand, the charging/discharging of the trap will modulate the maximum differential (ballistic) conductance through single transport channels [96].

At low temperatures Coulomb attractive traps have also been distinguished in CNT FETs [97], which exhibit opposite gate bias behavior of the amplitude, as shown in figure 4.24. A repulsive oxide trap is neutral before carrier capture and is

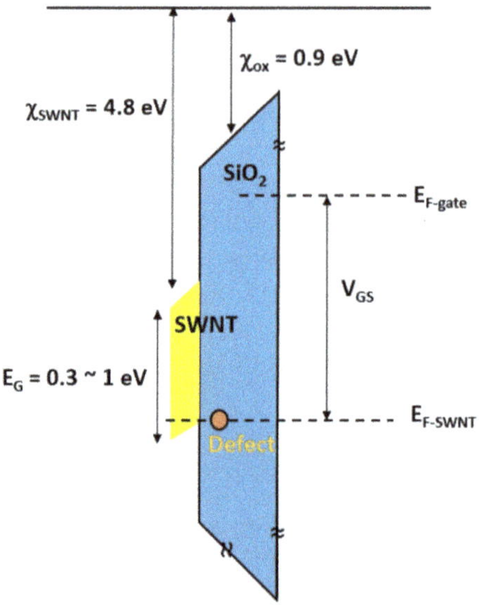

Figure 4.22. Energy band diagram illustrating the alignment of the Fermi energy with one defect level as the gate bias is varied. (After Liu *et al* [96].)

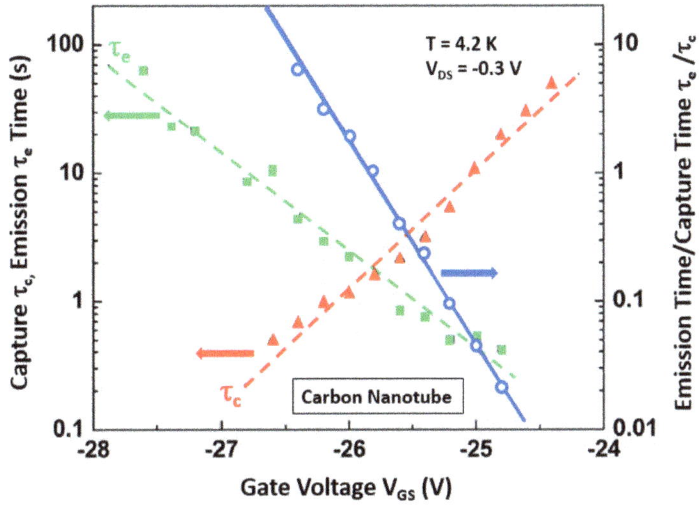

Figure 4.23. Statistics of emission/capture time constants and emission/capture time constant ratio for a single-defect level as a function of the gate bias. Emission/capture ratio follows detailed balance relation. (After Liu *et al* [96].)

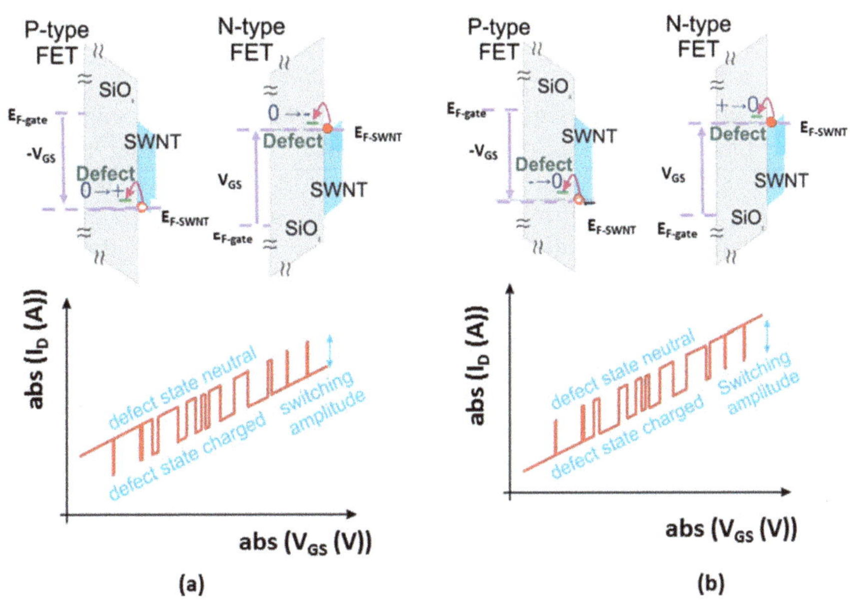

Figure 4.24. Band diagrams and schematic drawings of the corresponding I_D–V_{GS} relationships for a Coulomb repulsive RTS (*a*) and a Coulomb attractive RTS (*b*) in both p- and nFETs due to defects located near the CNTs. (Reproduced with permission from Liu *et al* [97]. Copyright 2006 the American Physical Society.)

characterized by a higher probability to be in a low current state for higher current. The carrier transitions can be described in terms of thermionic emission. This may no longer hold for an attractive trap which becomes neutral after majority carrier capture. The time constants in this case do not follow an exponential dependence on V_{GS} [97] and are more likely to be associated with a tunneling process, described by Fermi's golden rule (equation (2.25)).

It has also been observed that the RTS time constants of both attractive or repulsive traps do not exhibit a marked dependence on the magnitude of an applied magnetic field, either in the parallel or perpendicular direction [97]. Normally, the impact of a magnetic field is determined by the Zeeman energy E_Z, which modifies the detailed balance. For a repulsive trap, one expects that [97]

$$\frac{\tau_c}{\tau_e} = g\ \exp\left[\frac{E_T + E_Z - E_F}{kT}\right] \tag{4.8}$$

while

$$\frac{1}{\tau_e} = \left(\frac{2\pi}{\hbar}\right) D\Delta^2\left[1 - f(E_T)\right] \quad \text{and} \quad \frac{1}{\tau_c} = \left(\frac{2\pi}{\hbar}\right) D\Delta^2 f(E_T + E_Z) \tag{4.9}$$

for an attractive trap. It is believed that the RTS does not fit the Zeeman splitting model due to a broadening of the Zeeman levels at low temperature by scattering and spin relaxation [97]. Also, correlations between multiple-trap RTSs has been studied in CNT FETs [98, 99].

Recently, RTS has been investigated in metallic CNTs at 4.2 K [100], which behave differently than their semiconducting counterparts. An exponential dependence of the time constants on the drain-to-source bias was observed, which is different from standard behavior. With respect to the gate voltage dependence, it was noted that the peak positions of ΔI_D match those of the drain current [100]. At the same time, τ_c and τ_e did not vary with V_{GS}. The origin of these two-level fluctuators has been ascribed to defects inside the metallic CNT transferring between two metastable positions, which are separated by a barrier E_B, activated by inelastic scattering by the HC in the large longitudinal electric field. The reversible motion of the defect, inducing different electrical properties at each metastable position, could produce the observed telegraphic current fluctuations [100]. A model has been put forward which is similar to that for RTN in metallic nanobridges. It considers that the defect local temperature depends on V_{DS}, resulting in an exponential dependence of the time constants:

$$\tau = \tau_0 \exp\left(\frac{E_B - \xi V_{DS}}{\alpha_{CNT} q\, |V_{DS}|}\right) \tag{4.10}$$

In equation (4.10) τ_0 is an attempt period and ξ is an electromigration parameter. The experimental α_{CNT} is 0.034, which is one order smaller than the value of 5/6 observed for metallic nanobridges, perhaps due to the potential drop over the non-ideal ohmic contacts [100].

In the case of forward-biased nanotube film-silicon Schottky junctions, the observed RTSs have been ascribed to interface traps causing a modulation of the barrier height [101]. Finally, RTN has been found in CNT quantum nanodots at 4.2 K in the Coulomb blockade regime [102].

The importance of RTS in CNT devices has immediately triggered the interest of theorists [103–106]. Traps in the CNT and in the gate oxide have been considered, showing that for a charged impurity center in the channel a large scattering center is introduced [103]. It was also shown that a charge creates a larger V_T shift than a vacancy in the CNT. The latter creates a localized state in the band gap and a reduction of the transmission probability in both the CB and VB [103]. The effect of an impurity charge in the oxide is much smaller and decreases with distance to the interface. This is due to the screening effect of the gate electrode, which can be enhanced by replacing SiO_2 by a high-κ dielectric [103, 106]. In addition, it has been shown that a positive charge in the center of a CNT nFET creates a potential well in the CB, which attracts electrons, while a barrier is created for a negative charge [103, 104]. The opposite holds of course for a pFET. Overall, the impact of a charge creating a potential barrier will be much more pronounced than for a charge attracting carriers to a well, as quantum mechanical scattering from a well is weaker than from a barrier [103, 104]. Further studies have revealed that the effect of an oxide charge not only gives rise to a shift in V_T but also to a stretching along the V_{GS} scale if it resides very near the channel [104]. This is illustrated in figure 4.25. The normalized noise amplitude A_{RTS} is defined as the fractional current change due to

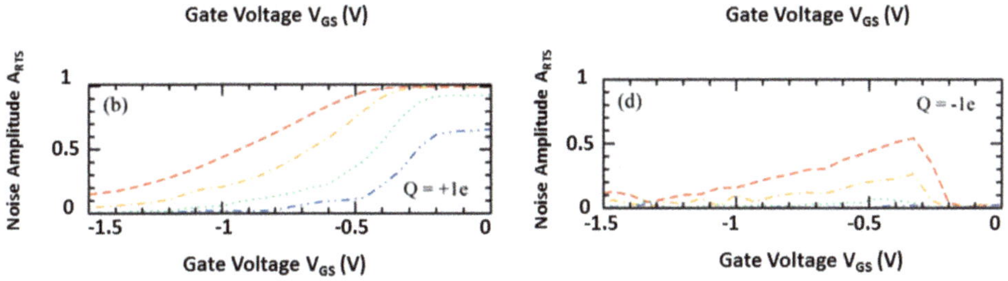

Figure 4.25. The effect of a trapped charge on current for different locations of the charge. Results are for a (13,0) CNT with $V_{DS} = 50$ mV, CNT diameter 32 nm and SiO_2 as a gate oxide. (*a*) Current versus gate voltage with no trapped charge (solid line) and for a positive charge at distances, from left to right of 0.4, 1.4, 3.4 and 7.4 nm from the nanotube wall. (*b*) Relative current change for data of (*a*). (*c*),(*d*) Same as (*a*),(*b*) for a negative charge in a CNT pFET. The insets in (*a*) and (*c*) show cross-sectional plots of the cylindrical geometry used in the calculations. (Reproduced with permission from Wang *et al* [104]. Copyright 2007 American Chemical Society.)

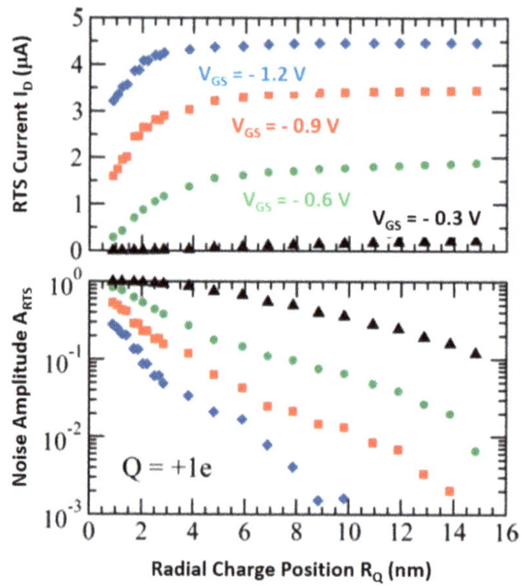

Figure 4.26. RTS current amplitude and noise versus radial charge position R_Q. Diamonds, squares, circles and triangles correspond to gate voltages $V_{GS} = -1.2$, -0.9, -0.6 and -0.3 V. $V_{DS} = 50$ mV and $d_{CNT} = 32$ nm. (Reproduced with permission from Wang *et al* [104]. Copyright 2007 American Chemical Society.)

the trapped charge, i.e. $A_{RTS} = (I_{D0} - I_{DT})/I_{D0}$ with I_{D0} and I_{DT} the current without and with the trapped charge. As can be seen in figure 4.25(*b*), for a positive charge A_{RTS} is lowest in the subthreshold regime, then increases with gate voltage and saturates when the transistor turns on. The sensitivity to a trapped charge can be substantially reduced by using a lower t_{ox} and a high-κ material [104, 106].

As shown in figure 4.26, A_{RTS} decreases in the first instance approximately exponentially with distance R_Q from the interface [104]. This can be understood in view of the tunneling transition through a barrier, which depends exponentially on the depth. The decay length $\partial R_Q/\partial \ln A_{RTS}$ becomes modified at full FET turn-on. At larger trap distances R_Q from the interface, the dependence is modified due to screening of the charge by the gate [104]. Also, the impact of the channel length on the RTS-induced ΔV_T has been calculated, showing that it increases for shorter FETs, considering a mid-channel oxide trap [105].

Finally, RTS has also been identified in organic materials, revealing deeper insight into transport mechanisms [107, 108].

4.4.2 Other advanced devices

Ultimately, devices handling single electrons or even spins will be employed for charge storage and sensing applications. This means the use of a 0D memory element, like a quantum dot or an SET. Given their small dimensions, the sensitivity to a nearby charge and associated RTS is expected to be high. Several reports have investigated RTS in silicon quantum dots [109, 110], confirming this idea. By their very high charge sensitivity, SETs are extremely sensitive to RTS, induced by a trap in close vicinity of the percolation channel [111, 112]. This has been exploited to propose a compact and low-power random-number generator for cryptographic applications [113]. The presence of a charge trap in an ET SOI film (~2 nm) close to the potential well of the SET (PP) gives rise to strongly amplified RTS with a low impact of thermal noise.

In nanoscale p–n junctions on SOI substrates, photon-induced RTS has been reported at liquid helium temperatures, which are ascribed to discharging in a donor–acceptor pair in the depletion region [114–116].

In addition to the transistor features, the ohmic contacts to source and drain should also scale accordingly. As recently shown, this introduces an RTS phenomenon in nanoscale contacts [117, 118], which is related to the Schottky barrier modulation by a trap in the vicinity of the nickel silicide/silicon (n-type) interface. The occurrence of such RTS is strongly dependent on the contact size and the technological details. It is assumed that the probability for a contact to contain, 1, 2, …, n RTSs follows a binomial Poisson distribution, given by [117, 118]

$$P_n = \frac{(N_t \times A)^n}{n!}\exp(-D_t A) \tag{4.11}$$

with N_t the uniform trap density in the silicon depletion region near the surface and $\sum_{n=0}^{\infty} P_n = 1$.

Finally, the occurrence of RTS in nanosystems employed for sensing applications has been demonstrated, for example, in [119, 120].

References

[1] Chim W K, Leong K K and Choi W K 2001 Random telegraphic signals and low-frequency noise in rapid-thermal-annealed silicon–silicon oxide structures *Jpn. J. Appl. Phys.* **40** 1–6

[2] Leyris C, Martinez F, Hoffmann A, Valenza M and Vildeuil J C 2007 N-MOSFET oxide trap characterization induced by nitridation process using RTS noise analysis *Microelectron. Reliab* **47** 41–5

[3] Sghaier N, Militaru L, Trabelsi M, Yacoubi N and Souifi A 2007 Analysis of slow traps centers in submicron MOSFETs by random telegraph signal technique *Microelectron. J.* **38** 610–4

[4] Lee H, Yoon Y, Jeon J and Shin H 2008 Analysis of random telegraph signal noise in dual and single oxide device and its application to complementary metal oxide semiconductor image sensor readout circuit *Jpn. J. Appl. Phys.* **47** 2602–5

[5] Chen M-J and Lu M-P 2002 On-off switching of edge direct tunneling currents in metal–oxide–semiconductor field-effect transistors *Appl. Phys. Lett.* **81** 3488–6

[6] Lee J-W, Lee B H, Shin H and Lee J-H 2010 Investigation of random telegraph noise in gate-induced drain leakage and gate edge direct tunneling currents of high-κ MOSFETs *IEEE Trans. Electron Devices* **57** 913–8

[7] Lee J-W, Park C H, Shin H, Park B-G and Lee J-H 2011 Accurate extraction of $\Delta I/I$ due to random telegraph noise in gate edge current of high-κ n-type metal–oxide–semiconductor field-effect transistors under accumulation mode *Appl. Phys. Lett.* **98** 023505

[8] Sakoda T *et al* 2010 Characterization of millisecond-anneal-induced defects in SiON and SiON/Si interface by gate current fluctuation measurement *Proc. Annual International Reliability Physics Symp.—IRPS10* (New York: IEEE) pp 379–84

[9] Tega N, Miki H, Mine T and Torii K 2012 Investigation of relationship between interface state and random telegraph noise using metal–oxide–semiconductor field-effect transistors fabricated on Si(100), (110), and (111) substrates *Jpn. J. Appl. Phys.* **51** 114001

[10] Yonezawa A, Teramoto A, Kuroda R, Suzuki H, Sugawa S and Ohmi T 2012 Statistical analysis of random telegraph noise reduction effect by separating channel from the interface *Proc. Annual International Reliability Physics Symp.—IRPS12* (New York: IEEE) pp 3B.5.1–3B5.7

[11] Chen C-E, Chang T-C, Lo H-P, Ho S-H, Lo W-L, Tseng T-Y, Cheng O and Huang C T 2012 Investigation of random telegraph signal with PD SOI MOSFETs *ECS Trans.* **45** 261–71

[12] Nishimura J, Saraya T and Hiramoto T 2011 Statistical comparison of random telegraph noise (RTN) in bulk and fully depleted SOI MOSFETs *Proc. 12th Int. Conf. on Ultimate Integration of Silicon (ULIS)* (New York: IEEE) pp 191–4

[13] Theodorou C G, Ionanidis E G, Haendler S, Planes N, Josse E, Dimitriadis C A and Ghibaudo G 2015 New LFN and RTN analysis methodology in 28 and 14 nm FD-SOI MOSFETs *Proc. Annual International Reliability Physics Symp.—IRPS15* (New York: IEEE) pp XT.1.1–1.6

[14] Nazarov A N, Ferain I, Dehdashti Akhavan N, Ravazi P, Yu R and Colinge J P 2011 Random telegraph-signal noise in junctionless transistors *Appl. Phys. Lett.* **98** 092111

[15] Nazarov A N, Lee C W, Kranti A, Ferain I, Yan R, Dehdashti Akhavan N, Ravazi P, Yu R and Colinge J P 2011 Comparative study of random telegraph noise in junctionless and inversion-mode MuGFETs *ECS Trans.* **35** 73–8

[16] Ferraton S, Militaru L, Souifi A, Monfray S and Skotnicki T 2008 Study of SiO_2/Si interface properties of SON MOSFETs by random telegraph signal and charge pumping measurements *Solid-State Electron.* **52** 44–8

[17] Trabelsi M, Militaru L, Sghaier N, Savio A, Monfray S and Souifi A 2011 Study of gate oxide/channel interface properties of SON MOSFETs by random telegraph signal and low frequency noise *IEEE Trans. Nanotech.* **10** 402–8

[18] Miki H, Yamaoka M, Frank D J, Cheng K, Park D-G, Leonbandung E and Torii K 2012 Voltage and temperature dependence of random telegraph noise in highly scaled HKMG ETSOI nFETs and its impact on logic delay uncertainty *2012 Symp. on VLSI Technol. Dig. of Techn. Papers* (New York: IEEE) pp 137–8

[19] Subirats A, Garros X, El Husseini J, Le Royer C, Reimbold G and Ghibaudo G 2013 Impact of single charge trapping on the variability of ultrascaled planar and trigate FD SOI MOSFETs: experiment versus simulation *IEEE Trans. Electron Devices* **60** 2604–10

[20] Fantini P and Ferrari G 2007 Low frequency noise and technology induced mechanical stress in MOSFETs *Microelectron. Reliab.* **47** 1218–21

[21] Lin M H, Hsieh E R, Chung S E, Tsai C H, Liu P W, Lin Y H, Tsai C T and Ma G H 2009 A new observation of strain-induced slow traps in advanced CMOS technology with process-induced strain using random telegraph noise measurement *2009 Symp. on VLSI Technol. Dig. of Techn. Papers* (New York: IEEE) pp 52–3

[22] Kwon H-M, Bok J-D, Han I-S, Park S-U, Jung Y J, Jang J-H, Ko S-Y, Lee W-M, Lee G-W and Lee H-D 2011 Dependence of hot carrier reliability and low frequency noise on channel stress in nanoscale n-channel metal–oxide–semiconductor field-effect transistors *Jpn. J. Appl. Phys.* **50** 10B0110B01

[23] Wu S L, Huang C W, Lu Y Y, Chang S J, Lin Y M, Lee K H and Cheng O 2012 Correlation between random telegraph noise and 1/f noise parameters in 28-nm pMOSFETs with tip-shaped SiGe source/drain *IEEE Electron Device Lett.* **33** 928–30
[24] Wang B-C, Wu S-L, Wu C-Y, Huang C-W, Lu Y-Y, Lin Y-M, Lee K-H, Cheng O, Huang P-C and Chang S-J 2012 Characterization of oxide traps in 28 nm nMOSFETs with different uniaxial tensile stress by utilizing random telegraph noise (RTN) *Extended Abstracts of the 2012 Int. Conf. on Solid State Devices and Materials* (Kyoto, Japan pp 114–5
[25] Wu S-L, Tsai K-S and Chang O 2013 Investigation of trap properties in high-κ/metal gate p-type metal–oxide–semiconductor field-effect transistors with SiGe source/drain using random telegraph noise analysis *Appl. Phys. Express* **6** 084201
[26] Wu S L, Tsai K-S and Cheng O 2013 Impact of SiGe source/drain induced compressive strain on low frequency noise in high-κ/metal gate p-channel metal–oxide–semiconductor transistors *Appl. Phys. Lett.* **103** 183508
[27] Huang P-C *et al* 2015 Impact of uniaxial strain in random telegraph noise in high-κ/metal gate pMOSFETs *IEEE Trans. Electron Devices* **62** 988–93
[28] Wang B-C, Wu S-L, Lu Y-Y, Huang C-W, Wu C-Y, Lin Y-M, Lee K-H, Cheng O, Huang P-C and Chang S-J 2013 Characterization of oxide traps in 28 nm n-type metal–oxide–semiconductor field-effect transistors with different uniaxial tensile stresses utilizing random telegraph noise *Jpn. J. Appl. Phys.* **52** 04CC24
[29] Veksler D, Bersuker G, Rumyantsev S, Shur M, Park H, Young C, Lim K Y, Taylor W and Jammy R 2010 Understanding noise measurements in MOSFETs: the role of traps structural relaxation *Proc. of the Int. Reliability Physics Symp.—IRPS10* (New York: IEEE) pp 73–9
[30] Cho H-J, Son Y, Oh B, Lee S, Lee J-H, Park B-G and Shin H 2010 Observation of slow oxide traps at MOSFETs having metal/high-κ gate dielectric stack in accumulation mode *IEEE Trans. Electron Devices* **57** 2697–703
[31] Lee J-W and Lee J-H 2012 Model of random telegraph noise in gate-induced drain leakage current of high-κ gate dielectric metal–oxide–semiconductor field-effect transistors *Appl. Phys. Lett.* **100** 033501
[32] Kobayashi S, Saitoh M and Uchida K 2008 ID fluctuations by stochastic single-hole trapping in high-κ dielectric p-MOSFETs *2008 Symp. VLSI Technol. Dig. Techn. Papers* (New York: IEEE) pp 78–9
[33] Tega N, Miki H, D'Emic C P, Zhu Y, Frank D J, Cai J, Guillorn M A, Park D-G, Haensch W and Torii K 2009 Reduction of random telegraph noise in high-κ/metal gate stacks for 22 nm generation FETs *Tech. Dig. Int. Electron Devices Meeting—IEDM09* (New York: IEEE) pp 771–4
[34] Shimizu T, Kato I, Yokogawa S, Kitagaki T, Taniguchi J, Suzuki T and Tsuboi T 2009 V_{TH} fluctuations due to random telegraph signal on work function control in Hf-doped silicate gate stack *Proc. 47th Annual International Reliability Physics Symp.—IRPS09* (New York: IEEE) pp 389–94
[35] Lee J-W, Lee B H, Shin H, Park B-G, Park Y J and Lee J-H 2010 Comparison of low-frequency noise in channel and gate-induced drain leakage current of high-κ nFETs *IEEE Electron Device Lett.* **31** 1086–8
[36] Kwon H-M *et al* 2011 Characterization of random telegraph signal noise of high-performance p-MOSFET with a high-κ dielectric/metal gate *IEEE Electron. Device Lett.* **32** 686–8

[37] Trabelsi M, Militaru L, Savio A, Monfray S and Souifi A 2011 High-κ gate stack properties in SON transistor given by voltage and temperature dependence of random telegraph signal *IEEE Trans. Electron Devices* **58** 1798–803

[38] Tega N, Miki H, Ren Z, D'Emic C P, Zhu Y, Frank D J, Guillorn M A, Park D-G, Haensch W and Torii K 2011 Impact of HK/MG stacks and future device scaling on RTN *Proc. Annual International Reliability Physics Symp.—IRPS11* (New York: IEEE) pp 630–5

[39] Tsai S C, Wu S L, Wu C-Y, Huang C W, Lu Y Y, Wang B C, Chang S J, Chen J F, Huang P C and Cheng O 2012 Impact of ZrO_2 on electron trapping behavior in 28 nm HKMG nMOSFETs by low-frequency (1/*f*) noise and RTN analysis *Extended Abstracts of the 2012 Int. Conf. on Solid State Devices and Materials* (Kyoto, Japan pp 120–1

[40] Wu S-L, Chiu H-F, Chang Y-S and Cheng O 2014 Defect properties of high-κ/metal-gate metal–oxide–semiconductor field-effect transistors determined by characterization of random telegraph noise *Jpn. J. Appl. Phys.* **53** 038005

[41] Chiu H F, Wu S L, Chang Y S, Chang S J, Chen J F, Tsai S C, Hsu C H, Lai C M, Hsu C W and Cheng O 2012 Impact of oxygen annealing on high-κ gate defects characterized by random telegraph noise *Appl. Phys. Lett.* **101** 122105

[42] Chiu H F, Wu S L, Chang Y S, Chang S J, Tsai S C, Huang P C and Cheng O 2012 Effect of annealing process on trap properties in metal/high-κ gate MOSFETs through RTN characterization *Extended Abstracts of the 2012 Int. Conf. on Solid State Devices and Materials* (Kyoto, Japan pp 112–3

[43] Chiu H F, Wu S L, Chang Y S, Chang S J, Huang P C, Chen J F, Tsai S C, Lai C M, Hsu C W and Cheng O 2013 Effect of annealing process on trap properties in high-κ/metal gate n-channel metal–oxide–semiconductor field-effect transistors through low-frequency noise and random telegraph noise characterization *Jpn. J. Appl. Phys.* **52** 04CC22

[44] Kao T-H, Chang S-J, Fang Y-K, Huang P-C, Lai C-M, Hsu C-W, Chen Y-W, Cheng O, Wu C-Y and Wu S-L 2014 Investigation of trap properties in high-κ/metal gate p-type metal–oxide–semiconductor field-effect-transistors with aluminum ion implantation using random telegraph noise analysis *Appl. Phys. Lett.* **105** 062109

[45] Kao T-H *et al* 2014 Impact of aluminum ion implantation on the low-frequency noise characteristics of Hf-based high-κ/metal gate pMOSFETs *IEEE Electron Device Lett.* **35** 954–6

[46] Kao T-H, Wu S-L, Tsai K-S, Fang Y-K, Lai C-M, Hsu C-W, Chen Y-W, Cheng O and Chang S-J 2014 Trap properties of high-κ/metal gate pMOSFETs with aluminum ion implantation by random telegraph noise and 1/*f* noise measurements *Jpn. J. Appl. Phys.* **53** 04EC14

[47] Chen C-E *et al* 2016 Analysis of oxide trap characteristics by random telegraph signals in nMOSFETs with HfO_2-based gate dielectrics *IEEE Electron Device Lett.* **37** 359–62

[48] Nagumo T, Takeuchi K, Hase T and Hayashi Y 2010 Statistical characterization of trap position, energy, amplitude and time constants by RTN measurement of multiple individual traps *Tech. Dig. Int. Electron Devices Meeting—IEDM10* (New York: IEEE) pp 628–31

[49] Realov S and Sheppard K L 2013 Analysis of random telegraph noise in 45-nm CMOS using on-chip characterization system *IEEE Trans. Electron Devices* **60** 1716–22

[50] Abe K, Teramoto A, Watabe S, Fujisawa T, Sugawa S, Kamata Y, Shibusawa K and Ohmi T 2010 Experimental investigation of effect of channel doping concentration on random telegraph signal noise *Jpn. J. Appl. Phys.* **49** 04DC07

[51] Kwon W H, Song Y H, Cai Y and Shim S P 2008 A novel structure for beyond-45 nm NOR Flash technology featuring short channel effect immunity and low random telegraph signal noise *Jpn. J. Appl. Phys.* **47** 8802–4
[52] Chen J, Higashi Y, Hirano I and Mitani Y 2013 Experimental study of channel doping concentration impacts on random telegraph signal noise and successful noise suppression by strain induced mobility enhancement *Symp. on VLSI Technol. Dig. of Techn. Papers* (New York: IEEE) T184–T185
[53] Abe K, Kumagai Y, Sugawa S, Watabe S, Teramoto A and Ohmi T 2009 Asymmetry of RTS characteristics along source–drain direction and statistical analysis of process-induced RTS *Proc. 47th Annual International Reliability Physics Symp.—IRPS09* (New York: IEEE) pp 996–1001
[54] Lukyanchikova N, Petrichuk M V, Simoen E and Claeys C 1997 Identification of isolation-edge related random telegraph signals in submicron silicon metal–oxide–semiconductor field-effect transistors *Appl. Phys. Lett.* **71** 3874–6
[55] Wang R-V, Lee Y-H, Lu Y-L R, McMahon W, Hu S and Ghetti A 2009 Shallow trench isolation edge effect on random telegraph signal noise and implications for Flash memory *IEEE Trans. Electron Devices* **56** 2107–13
[56] Chen J, Hirano I, Tatsumura K and Mitani Y 2012 Comprehensive investigations on neutral and attractive traps in random telegraph signal noise phenomena using (100)- and (110)-orientated CMOSFETs *Symp. on VLSI Technol. Dig. of Techn. Papers* (New York: IEEE) pp 141–2
[57] Schulz M 1993 Coulomb energy of traps in semiconductor space-charge regions *J. Appl. Phys.* **74** 2649–57
[58] Lee J W, Yun W S and Ghibaudo G 2014 Impact of trap localization on low-frequency noise in nanoscale device *J. Appl. Phys.* **115** 194501
[59] Cheung K P, Campbell J P, Potbhare S and Oates A 2012 The amplitude of random telegraph noise: scaling implications *Proc. Annual International Reliability Physics Symp.—IRPS12* (New York: IEEE) GD1.1–GD1.3
[60] Cheung K P, Campbell J P and Oates A 2011 High mobility channel from the prospective of random telegraph noise *Proc. Eur. Solid State Devices Conf.* (New York: IEEE) pp 335–8
[61] Amoroso S M, Monzio Compagnoni C, Ghetti A, Gerrer L, Spinelli A S, Lacaita A L and Asenov A 2013 Investigation of the RTN distribution of nanoscale MOS devices from subthreshold to on-state *IEEE Electron Device Lett.* **34** 683–5
[62] Amoroso S M, Georgiev V P, Gerrer L, Towie E, Wang X, Riddet C, Brown A R and Asenov A 2014 Inverse scaling trends for charge-trapping-induced degradation of FinFETs performance *IEEE Trans. Electron Devices* **61** 4014–8
[63] Pud S, Gasparyan F, Petrychuk M, Li J, Offenhäusser A and Vitusevich S A 2015 Single trap dynamics in electrolyte-gated Si-nanowire field effect transistor *J. Appl. Phys.* **115** 233705
[64] Gasparyan F, Zadorozhnyi I and Vitusevich S 2015 Single trap in liquid gated nanowire FETs: capture time behavior as a function of current *J. Appl. Phys.* **117** 174506
[65] Lim Y F *et al* 2006 Random telegraph signal noise in gate-all-around Si-FinFET with ultranarrow body *IEEE Electron Device Lett.* **27** 765–7
[66] Yang S, Yeo K H, Kim D-W, Seo K, Park D, Jin G, Oh K S and Shin H 2006 Random telegraph noise in n-type and p-type silicon nanowire transistors *Tech. Dig. Int. Electron Devices Meeting—IEDM08* (New York: IEEE)

[67] Zhuge J, Wang R, Huang R, Tian Y, Zhang L, Kim D-W, Park D and Wang Y 2009 Investigation of low-frequency noise in silicon nanowire MOSFETs *IEEE Electron Device Lett.* **30** 57–9

[68] Zhuge J, Zhang L, Wang R, Huang R, Kim D-W, Park D and Wang Y 2009 Random telegraph signal noise in gate-all-around silicon nanowire transistors featuring Coulomb-blockade characteristics *Appl. Phys. Lett.* **94** 083503

[69] Zhuge J, Wang R, Huang R, Zou J, Huang X, Kim D-W, Park D, Zhang X and Wang Y 2009 Experimental investigation and design optimization guidelines of characteristic variability in silicon nanowire CMOS technology *Tech. Dig. Int. Electron Devices Meeting—IEDM 09* (New York: IEEE) 61–4

[70] Zhang L, Zhuge J, Wang R, Huang R, Liu C, Wu D, Kang Z, Kim D-W, Park D and Wang Y 2009 New insights into oxide traps characterization in gate-all-around nanowire transistors with TiN metal gates based on combined I_g–I_d RTS technique *Symp. on VLSI Technol. Dig. Techn. Papers* (New York: IEEE) pp 46–7

[71] Liu C, Wang R, Zou J, Huang R, Fan C, Zhang L, Fan J, Ai Y and Wang Y 2011 New understanding of the statistics of random telegraph noise in Si nanowire transistors—the role of quantum confinement and non-stationary effects *Tech. Dig. Int. Electron Devices Meeting—IEDM11* (New York: IEEE) 521–4

[72] Saitoh M, Ota K, Tanaka C and Numata T 2013 Systematic understanding of channel-size dependence of low-frequency noise in 10 nm-diameter tri-gate nanowire MOSFETs *Symp. on VLSI Technol. Dig. Techn. Papers* (New York: IEEE) pp T228–T229

[73] Pud S, Li J, Petrichuk M, Feste S, Vitusevich S, Danilchenko B, Offenhäusser A and Mantl S 2013 Modulation phenomena in Si nanowire field-effect transistors characterized using noise spectroscopy and gamma radiation technique *J. Appl. Phys.* **113** 124503

[74] Vitusevich S 2015 Transport phenomena in liquid-gated Si nanowire FETs for biosensing applications *Proc. 2015 Int. Conf. on 1/f Noise and Fluctuations (ICNF)* (New York: IEEE) pp 1–4

[75] Ota K, Saitoh M, Tanaka C, Matsushita D and Numata T 2015 Experimental study of random telegraph noise in trigate nanowire MOSFETs *IEEE Trans. Electron Devices* **62** 3799–804

[76] Ohmori K, Feng W, Sato S, Hettiarachchi R, Sato M, Matsuki T, Kakushima K, Iwai H and Yamada K 2011 Direct real-time observation of channel potential fluctuation correlated to random telegraph noise of drain current using nanowire MOSFETs with four-probe terminals *Symp. on VLSI Technol. Dig. Techn. Papers* (New York: IEEE) pp 202–3

[77] Suzuki A, Kamioka T, Kamakura Y, Ohmori K, Yamada K and Watanabe T 2014 Source-induced RDF overwhelms RTN in nanowire transistor: statistical analysis with full device EMC/MD simulation accelerated by GPU computing *Tech. Dig. Int. Electron Devices Meeting—IEDM14* (New York: IEEE) pp 713–6

[78] Lee S-H, Kim Y-R, Hong J-H, Jeong E-Y, Yoon J-S, Baek C-K, Kim D-W, Lee J-S and Jeong Y-H 2014 Investigation of low-frequency noise in p-type nanowire FETs: effect of switched biasing condition and embedded SiGe layer *IEEE Electron Device Lett.* **35** 702–4

[79] Li J, Pud S, Petrychuk M, Offenhäusser A and Vitusevich S 2014 Sensitivity enhancement of Si nanowire field effect transistor biosensors using single trap phenomena *Nano Lett.* **14** 3504–9

[80] Clément N, Larrieu G, Nishiguchi K and Fujiwara A 2015 Ultra-low noise nanoscale transistors for metrology of noise, energy harvesting and biosensing *Proc. Int. Conf. on Noise and 1/f Fluctuations—ICNF 2015* (New York: IEEE) pp 1–6
[81] Lee T-H, Li Y-T and Hu S-F 2014 Random telegraph signal and spin characteristics of the gate-all-around poly-silicon nanowire *J. Appl. Phys.* **116** 174508
[82] Franco J *et al* 2012 Impact of single charged gate oxide defects on the performance and scaling of nanoscaled FETs *Proc. Annual International Reliability Physics Symp.—IRPS12* (New York: IEEE) pp 1–5
[83] Franco J, Kaczer B, Toledano-Luque M, Bukhori M F, Roussel Ph J, Grasser T, Asenov A and Groeseneken G 2012 Impact of individual charged gate-oxide defects on the entire I_D–V_G characteristics of nanoscaled FETs *IEEE Electron Device Lett.* **33** 779–81
[84] Franco J, Kaczer B, Toledano-Luque M, Roussel Ph J, Kauerauf T, Mitard J, Witters L, Grasser T and Groeseneken G 2013 SiGe channel technology: superior reliability toward ultra-thin EOT devices. Part II: Devices and other reliability issues *IEEE Trans. Electron Devices* **60** 405–12
[85] Toledano-Luque M, Kaczer B, Franco J, Roussel Ph J, Grasser T, Hoffmann T Y and Groeseneken G 2011 From mean values to distributions of BTI lifetime of deeply scaled FETs through atomistic understanding of the degradation *Int. Symp. on VLSI Technol. Dig. Techn. Papers* (New York: IEEE) pp 152–3
[86] Pogany D, Zeiner C, Bychikhin S, Burchhart T, Lugstein A and Vandamme L K J 2011 RTS and 1/*f* noise in Ge nanowire transistors *Proc. 21st Int. Conf on Noise and 1/f Fluctuations—ICNF 2011* (New York: IEEE) pp 368–71
[87] Conrad N, Si M, Shin S H, Gu J J, Zhang J, Alam M A and Ye P D 2014 Low-frequency noise and RTN on near-ballistic III–V GAA nanowire MOSFETs *Tech. Dig. Int. Electron Devices Meeting—IEDM14* (New York: IEEE) pp 502–5
[88] Si M, Shin S H, Conrad N J, Gu J, Zhang J, Alam M A and Ye P D 2015 Characterization and reliability of III–V gate-all-around MOSFETs *Proc. Annual International Reliability Physics Symp.—IRPS15* (New York: IEEE) pp 4A1.1–1.6
[89] Si M, Conrad N J, Shin S, Gu J, Zhang J, Alam M A and Ye P D 2015 Low-frequency noise and random telegraph noise on near-ballistic III–V MOSFETs *IEEE Trans. Electron Devices* **62** 3508–15
[90] Wan J, Le Royer C, Zaslavsky A and Cristoloveanu S 2010 Low-frequency noise behavior of tunneling field effect transistors *Appl. Phys. Lett.* **97** 243503
[91] Fan M-L, Hu V P-H, Chen Y-N, Su P and Chuang C-T 2013 Analysis of single-trap-induced random telegraph noise and its interaction with work function variation for tunnel FET *IEEE Electron Device Lett.* **60** 2038–44
[92] Huang Q, Huang R, Chen C, Wu C, Wang J, Wang C and Wang Y 2014 Deep insights into low frequency noise behavior of tunnel FETs with source junction engineering *Symp. on VLSI Technol. Dig. of Techn. Papers* (New York: IEEE) pp 88–9
[93] Liu F, Bao M, Kim H, Wang K L, Li C, Liu X and Zhou C 2005 Giant random telegraph signals in carbon nanotubes as a single defect probe *Appl. Phys. Lett.* **86** 163102
[94] Peng H B, Hughes M E and Golovchenko J A 2006 Room-temperature single charge sensitivity in carbon nanotube field-effect transistors *Appl. Phys. Lett.* **89** 243502
[95] Liu F, Wang K L, Zhang D and Zhou C 2006 Random telegraph signals and noise behavior in carbon nanotube transistors *Appl. Phys, Lett.* **89** 243101

[96] Liu F, Wang K L, Li C and Zhou C 2006 Study of random telegraph signals in single-walled carbon nanotube field effect transistors *IEEE Trans. Nanotechnol.* **5** 441–5
[97] Liu F, Bao M, Wang K L, Zhang D and Zhou C 2006 Coulomb attractive random telegraph signal in a single-walled carbon nanotube *Phys. Rev.* B **74** 035438
[98] Liu F and Wang K L 2008 Correlated random telegraph signal and low-frequency noise in carbon nanotube transistors *Nano Lett.* **8** 147–51
[99] Chan J, Burke B, Evans K, Williams K A, Vasudevan S, Liu M, Campbell J and Ghosh A W 2009 Reversal of current blockade in nanotube-based field effect transistors through multiple trap correlations *Phys. Rev.* B **80** 033402
[100] Chung H-J, Uhm T W, Kim S W, You Y G, Lee S W, Jhang S H, Campbell E E B and Park Y W 2014 Random telegraph noise in metallic single-walled carbon nanotubes *Appl. Phys. Lett.* **104** 193102
[101] An Y, Rao H, Bosman G and Ural A 2012 Random telegraph signal and 1/*f* noise in forward-biased single-walled carbon nanotube film-silicon Schottky junctions *Appl. Phys. Lett.* **100** 213102
[102] Jhang S J 2014 Analysis of random telegraph noise observed in semiconducting carbon nanotube quantum dots *Synthetic Metals* **198** 118–21
[103] Neophytou N, Kienle D, Polizzi E and Anantram M P 2006 Influence of defects on nanotube transistor performance *Appl. Phys. Lett.* **88** 242106
[104] Wang N-P, Heinze S and Tersoff J 2007 Random-telegraph-signal noise and device variability in ballistic nanotube transistors *Nano Lett.* **7** 910–3
[105] Wang N-P and Xu X-J 2013 Channel-length scaling for effects of single defects in carbon nanotube transistors *J. Appl. Phys.* **114** 073701
[106] Yu W-J and Wang N-P 2014 Effects of a single defect in composite gate insulators of carbon nanotube transistors *Eur. Phys. J.* B **87** 233
[107] Clément N, Pleutin S, Seitz O, Lenfant S and Vuillaume D 2007 $1/f^{\gamma}$ tunnel current noise through Si-bound alkyl monolayers *Phys. Rev.* B **76** 205407
[108] Brunner J, González M T, Schönenberger C and Calame M 2014 Random telegraph signals in molecular junctions *J. Phys.: Condens. Matter* **26** 474702
[109] Peters M G, Dijkhuis J I and Molenkamp L W 1999 Random telegraph signals and 1/*f* noise in a silicon quantum dot *J. Appl. Phys.* **86** 1523–6
[110] Kamioka J, Kodera T, Obata T, Tarucha S and Oda S 2014 Charge noise analysis of metal oxide semiconductor dual-gate Si/SiGe quantum point contacts *J. Appl. Phys.* **115** 203709
[111] Lo C C, Persaud A, Dhuey S, Olynick D, Borondics F, Martin M C, Bechtel H A, Bokor J and Schenkel T 2009 Device fabrication and transport measurements of FinFETs built with ^{28}Si SOI wafers toward donor qubits in silicon *Semicond. Sci. Technol.* **24** 105022
[112] Kano S, Azuma Y, Tanaka D, Sakamoto M, Teranishi T, Smith L W, Smith C G and Majima Y 2013 Random telegraph signals by alkanethiol-protected Au nanoparticles in chemically assembled single-electron transistors *J. Appl. Phys.* **114** 223717
[113] Uchida K, Tanamoto T and Fujita S 2007 Single-electron random-number generator (RNG) for highly secure ubiquitous computing applications *Solid-State Electron* **51** 1552–7
[114] Udhiarto A, Moraru D, Purwiyanti S, Kuzuya Y, Mizuno T, Mizuta H and Tabe M 2012 Photon-induced random telegraph signal due to potential fluctuations of a single donor-acceptor pair in nanoscale Si p–n junctions *Appl. Phys. Express* **5** 112201

[115] Purwiyanti S, Udhiarto A, Moraru D, Mizuno T and Tabe M 2012 Photoexcited-electron trapping by individual donor in lateral nanowire pn junction *Extended Abstracts of the 2012 Int. Conf. on Solid St. Devices and Materials* (Kyoto, Japan pp 945–6

[116] Purwiyanti S, Nowak R, Moraru D, Mizuno T, Hartanto D, Jablonski R and Tabe M 2013 Dopant-induced random telegraph signal in nanoscale lateral silicon pn diodes at low temperatures *Appl. Phys. Lett.* **103** 243102

[117] Chen M-C, Lin C-Y, Chen B-Y, Lin C-H, Huang G-W, Huang C-C, Ho C H, Wang T, Hu C and Yang F-L 2011 Silicide barrier engineering induced random telegraph noise in 1Xnm CMOS contacts *Tech. Dig. Int. Electron Devices Meeting—IEDM11* (New York: IEEE) 626–9

[118] Chen M-C, Lin C-Y, Chen B-Y, Lin C-H, Huang G-W, Ho C-H, Wang T, Hu C and Yang F-L 2012 Random telegraph noise in 1X-nm CMOS silicide contacts and a method to extract trap density *IEEE Electron Device Lett.* **33** 591–3

[119] Qiu H, Wang H and Ke F 2014 Low frequency noise in unstable contact region of Au-to-Au microcontact for microelectromechanical system switches *Appl. Phys. Lett.* **104** 254102

[120] Endean D E, Weigelt C T, Victoria R H and Dahlberg E D 2014 Tunable random telegraph noise in individual square permalloy dots *Appl. Phys. Lett.* **104** 252408

IOP Publishing

Random Telegraph Signals in Semiconductor Devices

Eddy Simoen and Cor Claeys

Chapter 5

Operational and reliability aspects of RTS

In the previous chapters, the effects of RTS on device performance have been studied under DC operating conditions. However, digital and analog circuits generally operate under alternating current (AC) bias at a certain AC modulation or switching. This means that the RTS should be studied under non-stationary switching conditions. As will be seen, this reduces the impact on device performance and can eventually suppress RTS. The opposite generally holds for high-field operation, which can create uniform Fowler–Nordheim or HC damage in the gate dielectric or at the interface. This leads to a change in the RTS behavior. The following section will focus on the connection between RTS and NBTI (or PBTI) in small-area transistors. It is shown that RTS constitutes the fundamental component of the charge relaxation at low V_{GS}, following BTI stress at high gate voltage (absolute value) and, therefore, of the so-called dynamic variability of scaled CMOS devices. In the final section, recent progress and efforts in the implementation of RTS-induced variability in device and circuit simulations are summarized.

5.1 Switching AC operation of RTS

In many analog and RF applications, a transistor does not operate under DC conditions, instead the voltage on the gate is switched between on and off (or from inversion to accumulation) under an AC voltage. As shown originally in [1], this has a strong impact on the RTS noise. In most cases, the noise amplitude is found to decrease, the reason being that the trap becomes saturated when biased into accumulation and no longer contributes to the fluctuations. This can be derived from the time constant behavior depicted in figure 5.1. More work along the same lines has been conducted by the Twente Group [2] confirming the initial observations. In addition, analytical modeling of the RTS noise under switched bias conditions has been reported [3–7]. Recently, a method has also been proposed to analyze the RTS parameters under cycled bias conditions [8]. Its principle is quite similar to the TDDS proposed by Grasser *et al* [9, 10]. It is concluded from these

doi:10.1088/978-0-7503-1272-1ch5

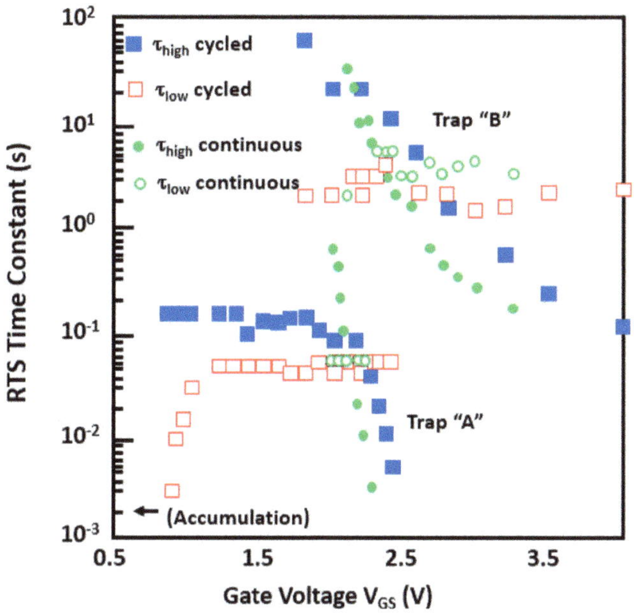

Figure 5.1. Time constants of two RTSs observed in a MOSFET switched between two gate voltages (V_{GS1} and V_{GS2}). τ_{high} and τ_{low} under 'continuous' operation ($V_{GS1} = V_{GS2}$ on the x-axis). τ_{high} and τ_{low} under 'cycling' operation ($V_{GS1} = 2.13$ V, V_{GS2} on the x-axis). (Reproduced with permission from Dierickx and Simoen [1]. Copyright 1992 AIP Publishing.)

studies that the trap state occupancy statistics do not instantly follow the applied voltage but rather depend on prior history. This reveals relevant information for the compact modeling of the impact of RTN on SRAM operation [8]. In the same context, the dependence of the RTS amplitude has been studied under switched forward substrate bias conditions, showing equally a decrease of its impact on the noise [11, 12]. It has been shown that the rising and falling edges of the pulsed AC gate voltage affect the RTS behavior as well [13, 14]. This has significant implications for the design of ultra-high speed circuits. A novel method has been developed for the analysis of multi-level AC RTS in scaled MOSFETs [15, 16]. The advantage is that it can catch the missing RTS statistics beyond the detectable window imposed by the gate voltage in the conventional DC RTS method. The impact of substrate biasing on RTS in digital circuits has also been addressed [17–20].

Along the same lines, the effect of microwave radiation on RTS noise has been studied [21, 22]. The main idea behind this investigation is that the use of RTS could be an enabling route to the read-out of qubits (single-electron detection) [23], by probing its spin state through microwave absorption [22]. A method has been proposed, based on electrically detected magnetic resonance, for the study of the oxide traps, which are responsible for gate-current RTS [24]. The detected electron spin resonance signal of an RTS trap in the pristine devices corresponds to the E'_γ center, originating from the O_3–Si dangling bond center. After hard BD (HBD), Si_3–Si traps were also observed in the SiO_2 using this technique [24]. Another interesting phenomenon is the occurrence of anomalous source current in accumulated pMOSFETs below 11.4 K [25], which is drastically enhanced with the application of a vertical magnetic field. At the same time, RTS has been found in this source current. It is believed that resonant tunneling processes between the source and the Fermi-edge singularity in the 2DEG at the channel cause these on/off fluctuations [25].

5.2 Impact of uniform and HC degradation

It is well-known that the gate current after oxide HBD exhibits irregular multi-level fluctuations [26]. TLFs occur already before fatal BD of the gate and are generally accepted as the fingerprint of defect creation and filamentary (non-uniform) current flow in a degraded dielectric. Overviews of the early work on this topic can be found in [27] and [28]. More recent studies have further focused on the physics of the post-HBD regime using RTS and LFN spectra [26, 29–34]. The resulting insights have led to the model represented in figure 5.2 [34] for digital RTS after HBD in thin gate oxides. Evidence has been provided that the traps involved in multiple trapping in a PP are oxygen vacancies [32, 33].

The physics-based model relies on the phonon TAT mechanism which is also used to explain stress-induced leakage current (SILC). The model allows in a unified way to explain the progressive degradation of the dielectric layer, without relying on different assumptions at different stages of the dielectric degradation process. According to figure 5.2, the HBD path consists of a large number of defects in a limited region, leading to a highly non-uniform trap distribution. This is usually accompanied by a local collapse of the oxide band gap [34]. The BD path is modeled through the three-shell picture of figure 5.2. Traps are randomly generated in the PP, resulting in a Gaussian profile along the radial direction of the cylinder. A uniform trap distribution is assumed along the depth (oxide thickness). A reduction of the

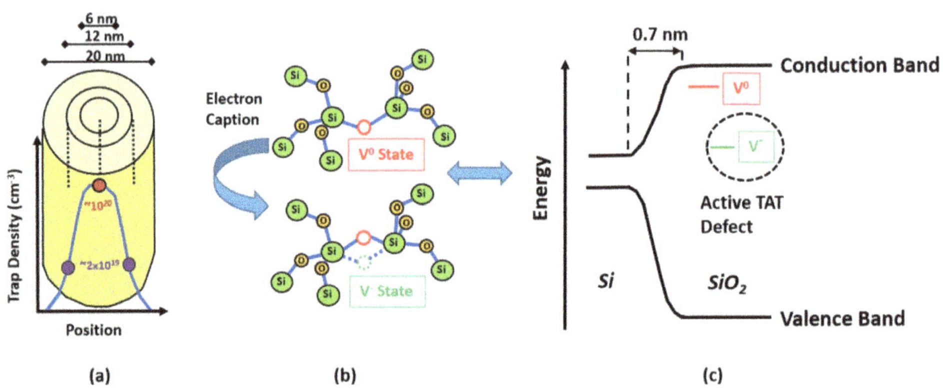

Figure 5.2. (*a*) Cylindrical three-shell structure used to model the BD path showing the Gaussian spatial profile for the random trap generation adopted along the radial direction. (*b*) Schematic representation of the switching trap model adopted to reproduce digital-BD I_G fluctuations. (*c*) Schematic band diagram showing the gradual transition between the Si and SiO_2 CB used also to account for the local CB lowering in the BD path. (Reproduced with permission from Padovani *et al* [34]. Copyright 2010 IEEE.)

oxide CB is implemented in the model, which is due to the assumed oxygen deficiency, similar to a transition layer between the SiO_2 and Si. RTS fluctuations then originate from oxygen vacancies which switch in the PP between two unstable states, associated with two atomistic oxygen vacancy configurations inside the SiO_2 network (V^0 and V^-), as depicted in figure 5.2 [34]. V^0 is a shallow trap ($E_T = 1.8$ to 2.6 eV above the Si CB), which corresponds to an inactive state with respect to the TAT conduction model. On the other hand, V^- is much deeper and better aligned with the Si CB, resulting in a much higher efficiency for electron capture. It thus corresponds to the active state in the TAT model. In this frame, electron capture corresponds to the turn-on of the defect and electron emission to its turn-off [34]. The model also explains why the relative amplitude of the gate-current fluctuations decreases with the number of traps in the PP and becomes higher for thinner oxides [34]. The model is stochastic in nature to reproduce the different spatial trap distributions and the gate voltage dependence. From the simulations it follows that the average number of oxygen vacancies is three to four in 1.6 nm SiON and five to six in 2.2 nm SiON [34].

Noise and fluctuation studies have also been reported in the quasi-BD regime [35–37]. RTS has been investigated in the gate current after soft BD (SBD) of HfSiON devices [38], showing drastic changes in the trap parameters, in particular the capture cross section and its barrier. LFN after channel soft oxide BD in HfLaSiO was reported quite recently [39], demonstrating that the addition of La leads to the creation of unstable La–O defect clusters close to the interface. In another study, the authors found the presence of both correlated and uncorrelated RTS in the gate and drain current after SBD (figure 5.3) [40]. They can have the same or opposite polarity. The occurrence of opposite polarity RTSs could be assigned to defects that become less effective to support TAT after an electron emission (e.g. E'_γ centers), while the I_D–I_G signals with positive polarity correspond

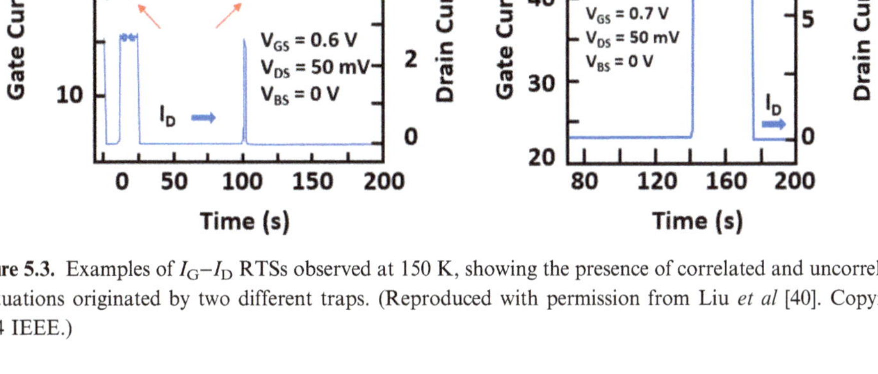

Figure 5.3. Examples of I_G–I_D RTSs observed at 150 K, showing the presence of correlated and uncorrelated fluctuations originated by two different traps. (Reproduced with permission from Liu *et al* [40]. Copyright 2014 IEEE.)

to the E' centers which become less effective TAT centers after electron capture [40]. It has been shown that the correlated RTSs correspond to traps in the SiO_2 IL close to the channel, while the uncorrelated signals have been assigned to defects in the high-κ layer [40]. In addition, RTS-like fluctuations have been observed after Fowler–Nordheim stress in the GIDL current of saddle type transistors for DRAM cells [41]. It was concluded that traps in the gate oxide can also be the origin of variable retention time.

A systematic RTS study after SBD/HBD of high-κ/metal gate transistors has been undertaken by the Nanyang Technical University of Singapore [42–44]. One of the important issues is to find out which layer breaks first: is it the interfacial oxide or the high-κ dielectric? The gate-current RTS was used to answer this question [42]. Based on the frequency exponent of the noise PSD and the time constants of the observed RTS, it was concluded that the high-κ layer breaks first under positive gate bias stress for the nMOSFETs studied [42]. Another curious observation was that the RTS amplitude and gate current decreased after bipolar switching, i.e. a positive V_{GS} constant voltage stress followed by a negative V_G sweep [43]. This indicates that a negative ramping switches the gate dielectric from a low- to a high-resistive state for the investigated NiSi fully silicided (FUSI)/HfSiON (2.5 nm)/SiO_2 (1.2 nm) and TiN/HfLaO (1.8 nm) gate stacks. This is illustrated in figure 5.4 [43]. It was concluded that faster RTS traps closer to the Si/SiO_2 interface or gate/SiO_2 interface are more easily 'passivated' by bipolar switching [43]. This is illustrated schematically by figure 5.5, assuming that oxygen vacancies are the source of RTS, while negative oxygen ions (O^-) drifting to the gate electrode act as a reservoir of passivating agents. The application of a negative gate voltage pushes the O^- back again in the oxide, where they can passivate the traps in the PP. This indicates a performance boosting technique [42]: application of a negative V_{GS} after some operation time to decrease the gate current and associated RTS of a degraded device.

Based on a systematic study of several BD events, a threshold criterion for the slope of the PSD has been established for the onset of RTS after BD, corresponding to a value of $\gamma = 1.5$ [45]. In the case of HfAlO, it was considered that the RTS traps (oxygen vacancies) are distributed in the interval 1.6–1.9 eV from the oxide CB and at 1.2–1.3 nm from the interface with silicon [45]. In addition, detailed studies have revealed that the PPs in the IL and high-κ layer are generally very close to each other, although different amounts of misalignment can occur, resulting in a lower SILC [46]. This can be advantageous from a reliability viewpoint.

The SILC at low gate bias is often characterized by RTS and some noise, and RTS studies have been reported [47–51]. In most cases, the metastable defect model by Farmer and Buhrman [27] is followed to interpret these data. In addition, it has been shown that there is little correlation between SILC and channel RTS in FN-stressed 5.7 nm SiO_2 gate nMOSFETs [49]. It was concluded that while both phenomena may originate from the same type of oxide defects, the ones responsible for RTS are closer to the interface. A detailed study of RTS in SILC reveals that the on and off time constants follow exponential distributions and decrease with the absolute value of the drain bias [50]. In addition, the off time constant is inversely

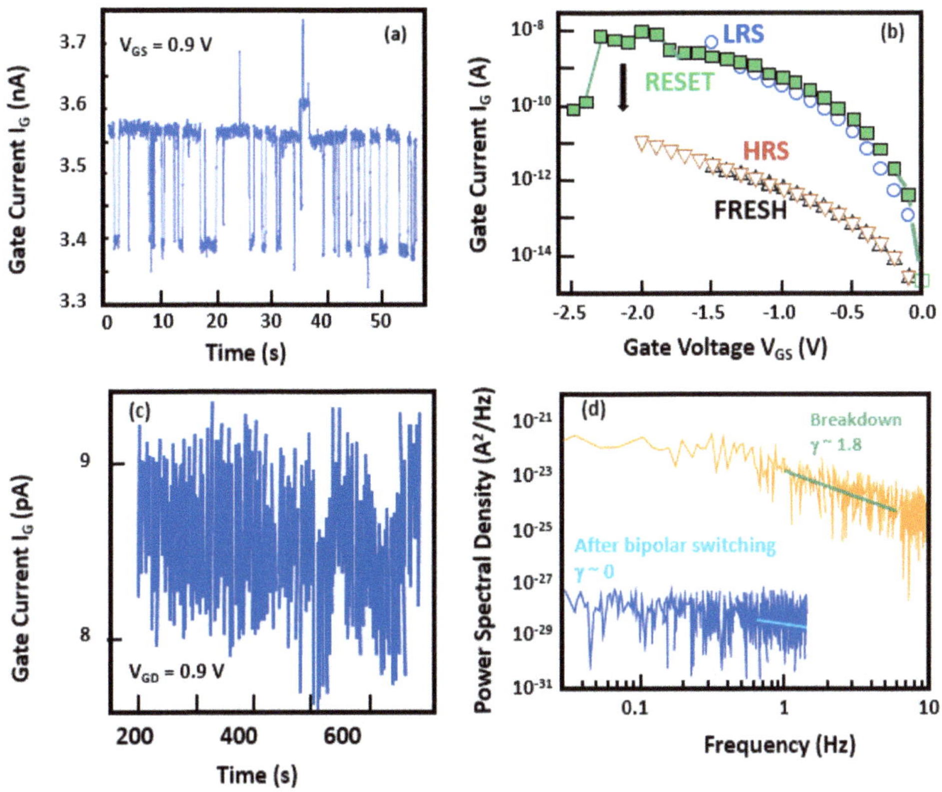

Figure 5.4. (*a*) Clear two-level I_G RTS at $V_{GS} = 0.9$ V, arising from the traps in the PP created from $I_G = 1$ μA BD in nMOSFET (NiSi FUSI/HfSiON/SiO_x and $W \times L = 1.0 \times 0.13$ μm^2). This indicates the existence of one active trap in the dielectric layer in the PP. (*b*) The $I_G - V_{GS}$ trend shows a bipolar switching after which I_G recovered fully to the fresh device condition. (*c*) At $V_{GD} = 0.9$ V after switching no RT signal was observed, in addition to a lower I_G level. (*d*) Comparison of I_G PSD between BD and after switching. The PSD measured at BD shows a slope of $\gamma = 1.8$, indicating the existence of Lorentzian RTS, while the one after switching is mainly attributed to white noise with $\gamma \sim 0$. There is also a strong decrease in the PSD magnitude. (Reproduced with permission from Liu *et al* [43]. Copyright 2011 IEEE.)

proportional to the total injected current ($I_G = I_{\text{dir-tunn}} + I_{\text{SILC}}$). τ_{on}, on the other hand, is not proportional to $1/I_G$ but to $1/I_{\text{SILC}}$.

Based on RTS in the SILC through the gate, a method has been proposed to observe the formation of the leakage path [52]. This is based on the observation that the traps created under gate voltage stress are distributed in energy around the Fermi level. The proximity of the leakage path determines the trap-induced gate voltage change ΔI_G, which can be used to determine the RTS distribution. The evolution of the gate-current path consists of three stages [52]. In the beginning, only direct-tunneling gate current and discrete oxide traps giving rise to I_G-RTS are observed. In the middle stage, interaction between traps and the PPs presents a multi-level gate-current variation. Finally, two different patterns for the HBD and SBD paths can be identified, which are illustrated in figure 5.6 [52]. The 'spindle' shapes of current paths are induced by a weaker condition of gate (BTI-) stress, while the

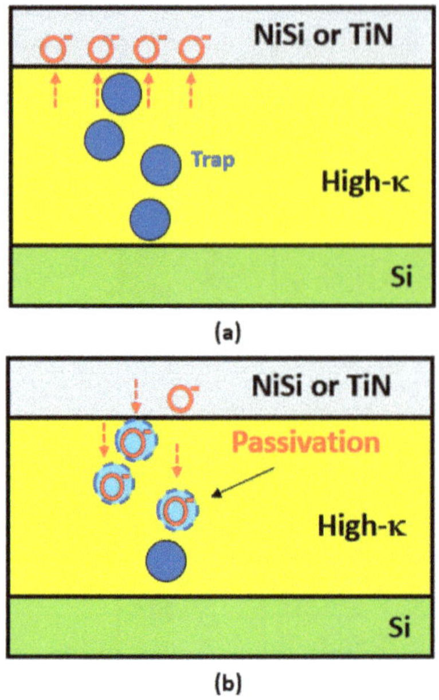

Figure 5.5. The reduction in RTN is modeled by moving oxygen ion passivating traps in the PP. Four traps in the dielectric layer corresponding to a 16-level RTS are reduced to a single-trap two-level RTS after passivation of three traps (*b*) from bipolar switching. (Reproduced with permission from Liu *et al* [43]. Copyright 2011 IEEE.)

'snake' shape is associated with a stronger stress condition, in terms of stress field (gate voltage) and temperature [52]. As a consequence, two different kinds of BD—HBD and SBD—can be triggered by the generation and location of traps and the interactions between traps and the gate-current paths. Actually, these two mechanisms can precisely explain the two-slope curves of time-to-BD prediction often observed in the literature [52].

Some investigations also focused on the impact of HC stress on RTS and the possible formation of new RTSs [53–55]. A general observation is that the character of stress-induced RTSs is different from that of the process-induced ones. This is illustrated by figure 5.7 [55]. Stress-induced traps are located closer to the interface and have faster switching times. A similar conclusion was reached for x-ray-induced RTSs [56]. Annealing at 400 °C in N_2 was able to remove most of the radiation-induced traps. In other investigations, it was much more difficult to reveal new stress-induced RTSs, possibly due to their fast time constants or the instable nature of the underlying traps [57, 58]. Instead, it was proposed to use the changes in pre-existing, process-induced RTSs as a local probe for the HC-induced damage at the interface and in the nearby oxide. Recently, RTS in the gate leakage current has been used to study HC stress-induced defects in 2.6 nm SiO_2/poly-Si 80 nm DRAM technology [59].

Similar to the BTI phenomenon, there is a quantization (or discretization) of HC degradation in small-area MOSFETs [60]. The degradation occurs stepwise, by

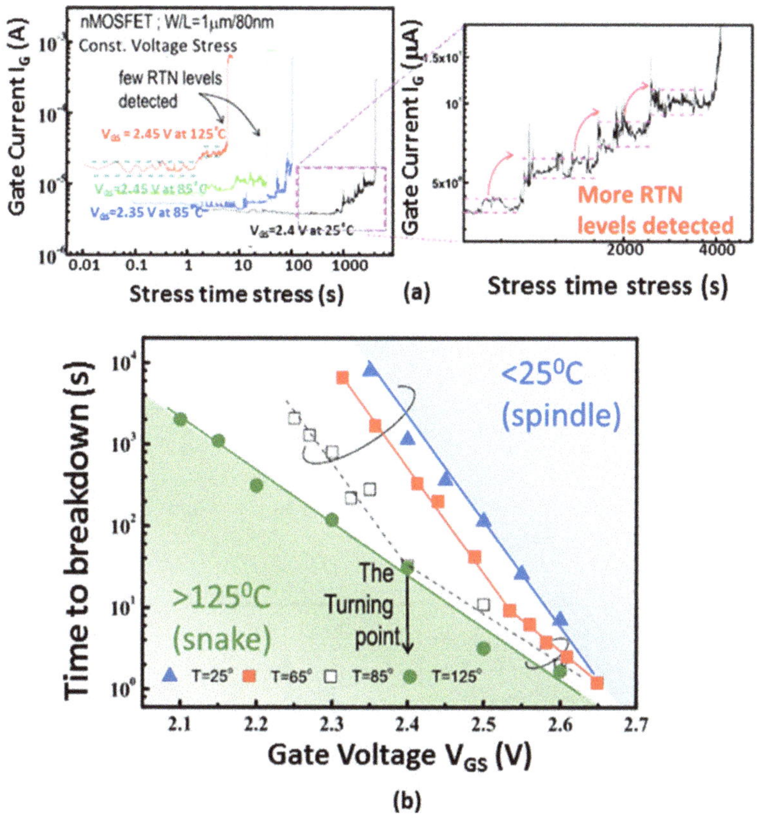

Figure 5.6. (*a*) The transient characteristics of two different types of leakage current. One (black) shows many levels of gate current allowing for more SBD to happen, leading to a longer time-to-BD. The other (red) exhibits fewer levels of gate current with shorter time-to-BD. (*b*) The experimental results in the prediction of time-to-BD lifetime. A two-slope curve has been identified. The temperature and intensity of the stressing electric field dominate the magnitude of the induced gate currents. (Reproduced with permission from Hsieh and Chung [52]. Copyright 2015 AIP Publishing.)

single-trap-induced degradation (STID). This STID follows a Weibull distribution and scales with 1/Area, irrespective of the technology (SiON/poly, high-κ/metal gate or FF) [60]. From the comparison in figure 5.8 of different single-trap-related degradation modes, it appears that the impact of HC injection (HCI) in an nMOSFET is larger than NBTI (pMOS) or normal RTN behavior.

In NWFETs, HCI (and NBTI) leads to higher RTS amplitudes, particularly in smaller diameter NWs, which is related to traps created at the channel corner [61]. A similar tendency has been found after inductively coupled plasma (ICP) treatment of both SiO_2 and high-κ/poly-MOSFETs [62]. It shows that the RTS amplitude can be a useful parameter to study the ICP damage, associated with the etching of Al metallization (the antenna effect).

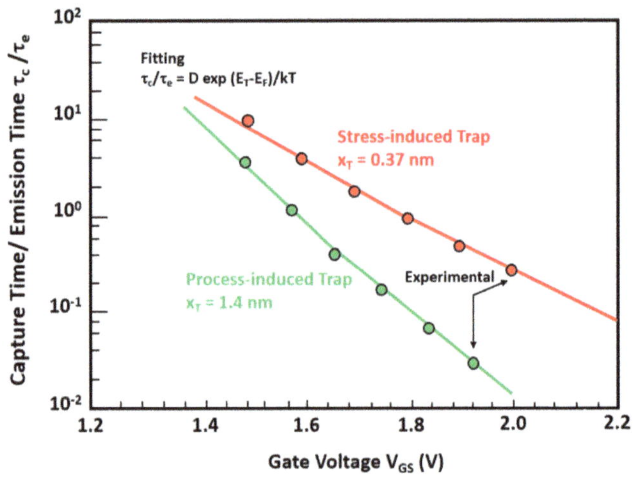

Figure 5.7. τ_c/τ_e of process-induced and stress-induced traps versus gate bias. The trap location x_T can be determined from this plot. (After Fang *et al* [55].)

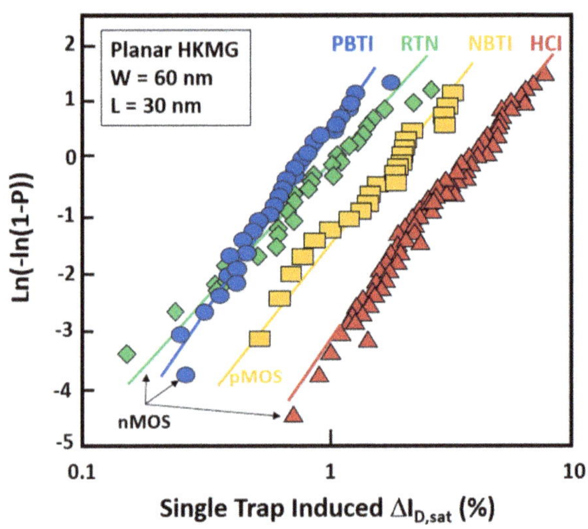

Figure 5.8. The comparison of STID in HCI, NBTI, PBTI and RTN. The results imply the defect locations of the different degradation mechanisms, since the STID impact can be larger when the trap is located near the channel. (After Liu *et al* [60]).

5.3 BTI and RTS: oxide trapping?

In recent years, the scaling to the deep submicron regime has caused a paradigm shift in the characterization and modeling of small-area transistors, which is related to the PV and, in particular, the RDFs. Due to the statistical nature of many process steps (ion implantation, dielectric layer deposition, etching and lithography, metal gate deposition, etc) a certain variation in the device parameters occurs over the wafer and from wafer to wafer, which has to be taken into account in device and circuit modeling. With some years of delay, the same shift from deterministic to stochastic is taking place in the description of the reliability of deep submicron CMOS components [63]. This is perhaps best illustrated with the degradation and relaxation associated with BTI (discussed in the next paragraph). The reliability of nanoscale MOSFETs has become a statistical problem, characterized by a statistical distribution with relatively rare but acute changes in the device parameters due to charge trapping in the fluctuating potential landscape generated by random discrete dopants and other statistical variability sources.

The analogy between NBTI relaxation and low-frequency 1/f noise in pMOSFETs was firmly established a few years ago [63, 64]. In particular the relaxation transient at low gate bias following BTI stress shows a strong analogy with the distribution of time constants which are required to explain the $1/f$ noise based on the McWhorter tunneling-into-oxide-traps model. This has led to the important conclusion that the reversible (or recoverable) part of BTI is due to the charging of the same type of oxide traps which are also responsible for the $1/f$ noise. The analogy goes even further when studying relaxation transients in submicron area transistors, as shown in figure 5.9 [65]. One can clearly see that the relaxation transient is no longer a smooth curve but breaks down in discrete steps as in figure 2.7 [66, 67], which are due to the random emission of a trapped electron (nMOSFET) or hole (pMOSFET) from various traps in the oxide. As in RTS, the different traps can be identified by the ΔV_T step height (or RTS amplitude), while the release (emission) time is randomly distributed, following exponential statistics [65]. This has initiated more

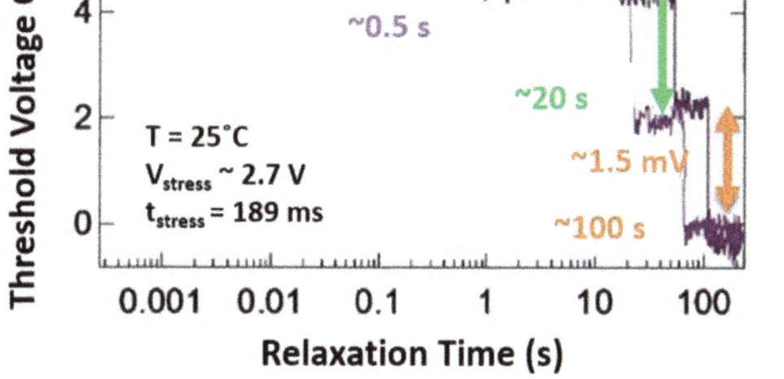

Figure 5.9. Characteristic ΔV_T transients of a single 70 nm × 90 nm nFET with a 1 nm SiO_2/1.8 nm HfSiO gate stack, stressed at 25 °C and V_{GS} = 2.7 V for 189 ms. Four discrete drops are observed, indicating the existence of at least four active traps at the stress condition. (Reproduced with permission from Kaczer *et al* [65]. Copyright 2011 AIP Publishing.)

studies on RTS following NBTI and PBTI. For an overview of the current state of affairs, the reader is referred to the work of Grasser [68].

Figure 5.9 also gives the measurement principle of the so-called TDDS recently introduced by Grasser *et al* [9, 10, 69]. It can be considered as an extension of the DLTS technique [57, 67], whereby spectral maps are used to identify the individual defects. Such a map is represented in figure 5.10 [65] and plots the trap amplitude (ΔV_T) versus the emission time constant (τ_e), for a fixed stress time (charging or trap filling time). In contrast to standard RTS measurements, the device is switched to a lower gate voltage after a charging period at a higher V_{GS} for a specific time t_s. So, while the first type of measurement is in dynamic equilibrium (or steady-state), the TDDS technique probes the RTS properties in a relaxation transient towards steady-state. By changing the stress time one also changes the filling of the traps and the probability of emission during the relaxation period. In other words, the number of hits per cluster in the spectral maps will change exponentially according to [9]

$$P_c^{(i)} = 1 - \exp\left(-\frac{t_s}{\tau_c^{(i)}}\right) \tag{5.1}$$

Thus, recording M spectral maps with increasing stress time enables one to determine τ_c *at the higher stress gate bias* [9, 10]. In this way, both the bias and temperature dependence of τ_e and τ_c can be studied, revealing a thermally activated

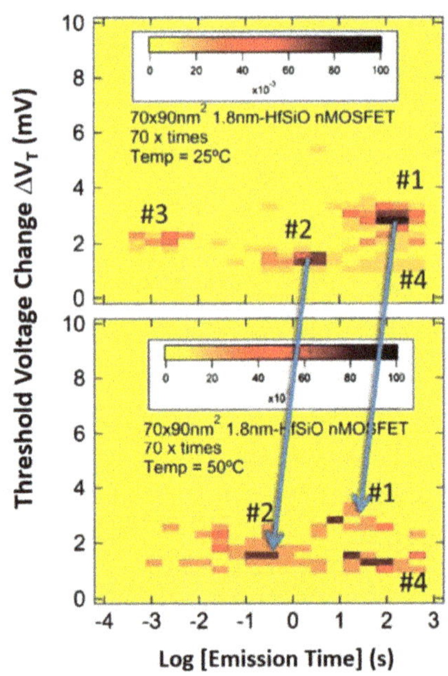

Figure 5.10. Two-dimensional histograms (TDDS spectra) of the heights and emission times of the steps extracted from 70 ΔV_T transients of the device of figure 5.8 at (*a*) 25 °C and (*b*) 50 °C. Four clusters are formed that shift horizontally to shorter emission times with increasing temperature. Note that trap 3 disappears from the experimental window at 50 °C. (Reproduced with permission from Kaczer *et al* [65]. Copyright 2011 AIP Publishing.)

behavior [9, 10, 65], as can be seen in figure 5.10. The same technique can be used to study the impact of AC versus DC stress on the trap occupation statistics [70] in order to understand the implications for BTI under cycling operation.

One important difference between normal RTN in dynamic equilibrium and NBTI RTS is that the range of active traps in energy and distance from the interface is much larger in the latter case [71]. In order to observe RTS in a transistor, one should have that E_T is close to E_F or in other words $\tau_c \sim \tau_e$, so that only a rather narrow energy interval of a few kT around E_F is accessible. In the case of BTI, the active window is much wider, as τ_e can be very different from τ_c. The active region is then mainly determined by the field during stress [68], as schematically shown in figure 5.11.

Based on these TDDS studies a firm link has been established between BTI and RTS in submicron MOSFETs [65, 68–70, 72–75] and a dedicated switching trap model has been proposed to explain the gate bias dependence of the RTS time constants [10, 65, 68–70, 72–74], which resembles the model for HBD RTS of figure 5.2. In addition, a statistical model for the BTI distribution has been established, which takes into account the random (Poisson) distribution of the number of traps in the device and the exponential distribution of trap amplitudes ΔV_T, which is illustrated by figure 5.12 [65]. The total ΔV_T distribution is thus a summation of the ΔV_T cumulative distribution function weighted by the Poisson probability of the number of traps per device [65]. In this way, satisfactory agreement is obtained between experimental and simulated distributions of BTI-induced V_T shifts [65]. In addition, it has been demonstrated that the V_T degradation under NBTI evolves according to log(t), in agreement with the trapping–detrapping view rather than with the exponential t^n law (n ~ 0.1–0.3), assumed in the past [76]. The former model correctly reproduces the aging variability due to the randomness in number of available traps, and accurately predicts the mean and the variance of the V_T shift. On the other hand, the traditional reaction–diffusion based t^n law overestimates the aging rate and exaggerates its variance [76]. Based on a critical discussion of the two types of theories for BTI, a modeling benchmark has been proposed, relying on the degradation behavior of nanoscale MOSFETs [77].

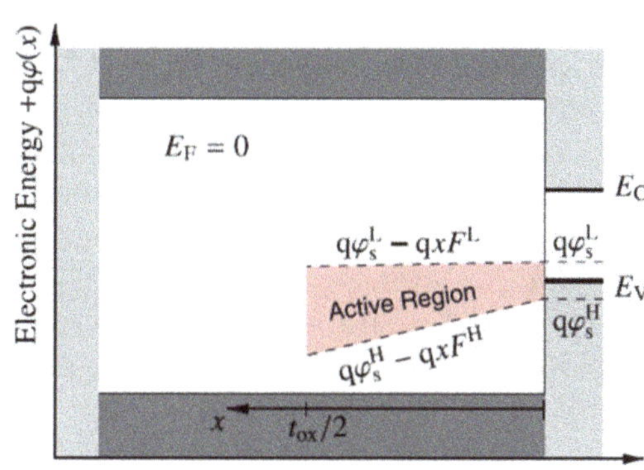

Figure 5.11. The active region of BTI depends dominantly on the field during stress. Only defects in the active region are neutral prior to stress and can be potentially charged during stress. (After Grasser [69].)

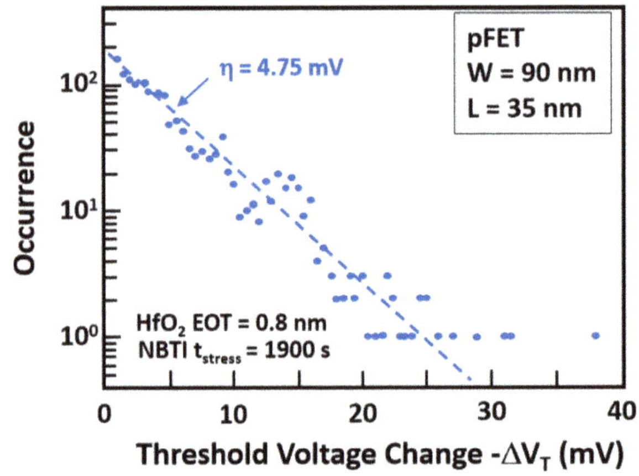

Figure 5.12. Histogram of NBTI transient individual step heights measured on 72 devices showing a clear exponential distribution. The average V_T shift η corresponding to a single carrier discharge is 4.75 ± 0.30 mV in the pFETs with metallurgical length L = 35 nm, W = 90 nm and a HfO_2 dielectric with EOT = 0.8 nm. (Reproduced with permission from Kaczer *et al* [65]. Copyright 1992 AIP Publishing.)

The usual Poisson model for the distribution of the number of traps has recently been challenged by Chiu *et al* [78], arguing that the newly created traps under BTI stress are not necessarily independent of the pre-existing traps. From the NBTI trap growth rate with $t^{1/6}$ it is derived that the probability for a new trap being created decreases with an increasing number of trapped charges. This implies that the use of the Poisson distribution may overestimate the total ΔV_T distribution [78]. Recent 3D atomistic simulations lend support for this point of view, showing that the mutual interaction is fundamental in determining the dispersion of capture and emission time constants in BTI simulations [79]. This stands in contrast to the standard single-defect RTS observed in the steady-state, which typically follows the criterion of $\tau_c/\tau_e \sim 1$, selecting traps closer to the interface that have a larger impact on ΔV_T. In this case, the change induced by a single charge is hardly affected by the presence of other traps in the oxide, i.e. there is only a small mutual interaction [79].

Relying on the new insights provided by the trapping–detrapping approach of BTI, a revised model has been established for large-area transistors, which has shown that it is useful to rely on so-called capture/emission time (CEMT) maps [63, 80, 81]. Such a CEMT map provides a convenient transformation of the experimental data to help understand the wide distribution of time constants. It also explains the frequency dependence of recoverable damage under dynamic stress [81].

The correlation between RTN and NBTI-stress-induced traps has been further investigated in the frame of the as-grown-generation model [82]. Clear evidence has been presented that the as-grown hole traps (AHTs) are responsible for the RTS of pMOSFETs. AHTs also dominate NBTI initially, but the generated defects (GDs) become increasingly important with progressing stress time. The GDs, however, do not generate RTS [82].

As mentioned above, AC operation has a strong impact on the LFN and on RTS. This has also been investigated for the case of NBTI [83] and PBTI [84]. In the first case, the AC stress response or cycling of individual RTSs in SiON pMOSFETs has been studied, whereby switching from the stress gate voltage V_{stress} to the relaxation gate voltage V_{relax} was applied at a frequency of 100 kHz with a duty factor (DF) = $(f \times t_h \times 100)$. The time t_h is when the device is at the high stress bias for a total stress time of t_{stress} [83]. The probability of hole capture during a DC stress (for DF = 1) can be written as

$$P_c(t_{stress}) = \frac{\tau_e}{\tau_e + \tau_c}\left\{1 - \exp\left[-\left(\frac{1}{\tau_e} + \frac{1}{\tau_c}\right)t_{stress}\right]\right\} \tag{5.2}$$

and

$$P_e(t_{relax}) = \frac{\tau_c}{\tau_e + \tau_c}\left\{1 - \exp\left[-\left(\frac{1}{\tau_e} + \frac{1}{\tau_c}\right)t_{relax}\right]\right\} \tag{5.3}$$

keeping in mind that τ_e is dependent on V_{GS}. In addition, one should take into account that τ_c decreases with increasing $|V_{stress}|$, while τ_e normally increases.

According to figure 5.13, the occupancy probability of equations (5.2) and (5.3) increases with DF up to the DC value, as expected. This can be explained by considering that the number of traces not containing any hole capture event decreases with smaller DF, i.e. the time that the voltage is at the 'high' level [83].

The occupancy probability for the case of digital switching stress has been worked out and is calculated in figure 5.14. It shows that the higher the DF is, the higher the occupancy probability. This follows from the fact that the occupancy increases during the high gate voltage period and decreases during the relaxation time. If a number n of stress pulses is applied ($n = f \times t_{stress}$), the occupancy can be written as [83]

$$P_c(n) = \frac{b}{a}[1 - \exp(-an)] \tag{5.4}$$

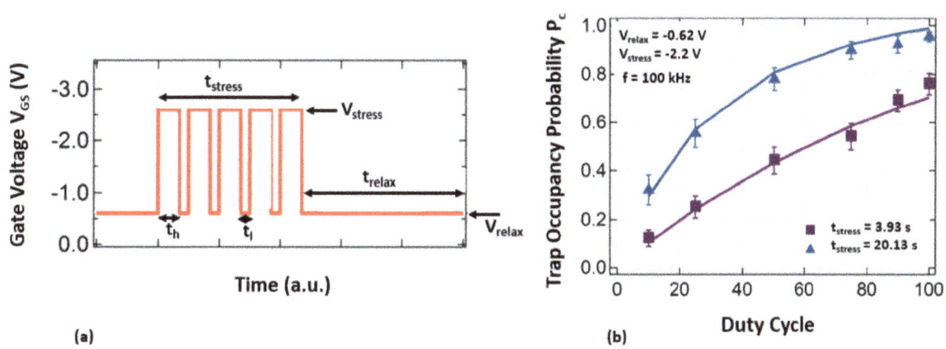

Figure 5.13. (*a*) The applied AC signal switches between V_{stress} and V_{relax} for a stress time t_{stress}. Afterwards, the relaxation transient is registered at V_{relax} for a relaxation time t_{relax}. (*b*) (Symbol) Trap occupancy probability P_c (number of traces that presents a giant step/number of trials). (Lines) Predicted occupancy according to the proposed theory considering the emission and capture times at V_{stress} = −2.2 V and V_{relax} = −0.62 V. (Reproduced with permission from Toledano *et al* [83]. Copyright 2011 IEEE.)

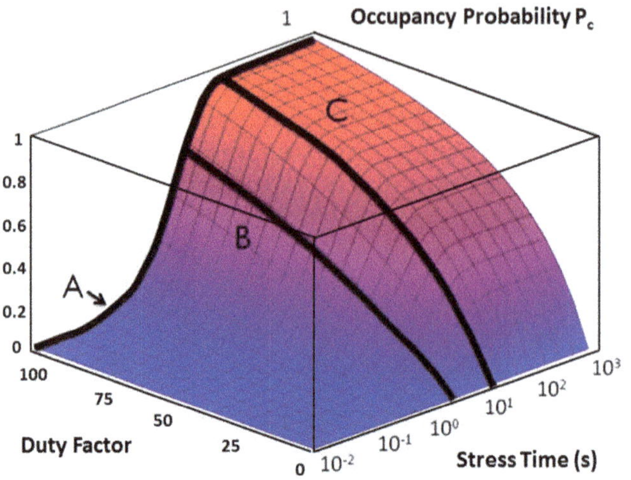

Figure 5.14. Occupancy probability P_c as a function of t_{stress} and DF. Line A traces P_c versus t_{stress} at DC stress. Lines B and C trace P_c versus DF. (Reproduced with permission from Toledano *et al* [83]. Copyright 2011 IEEE.)

which reduces to a simpler form if the characteristic emission and capture times are significantly larger than the period of the stress signal

$$P_c(t_{stress}) = \frac{\tau_e^*}{\tau_e^* + \tau_c^*}\left(1 - \exp\left[-\left(\frac{1}{\tau_c^*} + \frac{1}{\tau_e^*}\right)t_{stress}\right]\right) \tag{5.5}$$

where τ_c^* and τ_e^* are the effective capture and emission time under AC stress which depend on the capture and emission times at V_{stress} and V_{relax}, respectively [83]. They are given by

$$\frac{1}{\tau_c^*} = \frac{\text{DF}}{\tau_c V_{stress}} + \frac{1 - \text{DF}}{\tau_c V_{relax}} \tag{5.6}$$

and

$$\frac{1}{\tau_e^*} = \frac{\text{DF}}{\tau_e V_{stress}} + \frac{1 - \text{DF}}{\tau_e V_{relax}} \tag{5.7}$$

It can be seen from equations (5.5)–(5.7) that the occupancy probability increases with V_{stress}, with t_{stress} and with the DF (figure 5.14). The model can correctly predict the P_c for all t_{stress} and DF provided that the emission and capture times under DC stress conditions are known.

The effect of PBTI/NBTI on I_D-RTS [85–88] and I_G-RTS has been studied on metal gate/high-κ transistors [89, 90], showing among others a marked drop in amplitude compared with polysilicon/SiO_2 gate stacks. Traps newly created by the stress have also been identified in this way. A clear correlation was reported between RTS in I_G and I_D in high-κ MOSFETs [91], whereby the I_G-RTN is directly related to the physical trapping or detrapping and the I_D-RTS reflects the sensitivity to charge trapping as determined by g_m. The occurrence of RTS-induced by NBTI or PBTI is also highly sensitive to the gate stack processing, e.g. the nitrogen distribution in nitrided gate oxides is an important parameter [92]. It has been demonstrated that the worst case bias condition for observing RTS in nanoscale pMOS devices usually occurs at high gate voltage and low drain voltage (linear operation) but it may change under dynamic operation [93]. At the same time, the RTS amplitude after BTI stress remains at the same range as under the 'time zero' condition [93]. A systematic comparison between NBTI/pMOS and PBTI/nMOS has been reported for SiON transistors, relying on TDDS analysis [94]. Overall, a ten times lower trap density has been revealed in nMOS devices, with amplitudes following an exponential distribution [94]. BTI has also been studied in the inversion and accumulation modes of high-κ/metal gate 28 nm CMOS devices, allowing the detection of traps in a wide energy range in the high-κ dielectric [95]. The inversion mode measurement can only identify RTN traps in the channel region, while in accumulation mode traps inside the gate–drain overlap region are accessible—these are also at the origin of the GIDL current.

In recent years, BTI studies of forthcoming or exploratory nanoscale technologies have been reported, comprising RMG-high-κ/metal gate FFs [96], NWFETs [97] and InGaAs FFs [98]. The observation of RTS in aged AlGaN/GaN high electron-mobility transistors has occasionally been reported as well [99]. A systematic study of the impact of HC and BTI stress on tri-gate SNWTs has revealed the picture summarized in figure 5.15 [100]. The RTS increases for narrower NW diameters due to the higher electric field in the corner. The RTS amplitude increase with HC stress time. This means that more traps are created with an average higher ΔV_T amplitude closer to the interface and in the vicinity of the minimum potential path (critical current path) [100]. A similar conclusion was reached after BTI stress.

5.4 Statistical RTS measurement methods

As RTS is a rather rare event in present-day CMOS technologies, a large data set is required in order to enable a statistical assessment of its impact on the V_T fluctuations and to address the issue of rare but extreme values. Therefore, dedicated measurement methods have been devised employing special test structures [101–104] or a read-out of a memory array [105–108]. An example of a test structure suitable for evaluating the threshold voltage of a large number of transistors is represented in figure 5.16 and the measurement principle is outlined in figure 5.17 [103]. In addition, the large set of data also requires a dedicated analysis scheme and data handling [8, 109–111]. An efficient evaluation of the distribution of the RTS time constants has been proposed by Obara *et al* [112]. A novel statistical analysis method of the dynamic variability, induced by RTS, has been proposed for 28 nm FD-SOI MOSFETs [113, 114]. It was concluded that the dynamic variability is reduced for increasing device area and increased measurement rise time t_{rise} of the gate voltage ramp during the V_T measurement on a MOSFET. An area dependence, as for static mismatch, has been derived, scaling with $1/\sqrt{WL}$. An expression for the V_T dynamic matching parameter was established, given by [113]

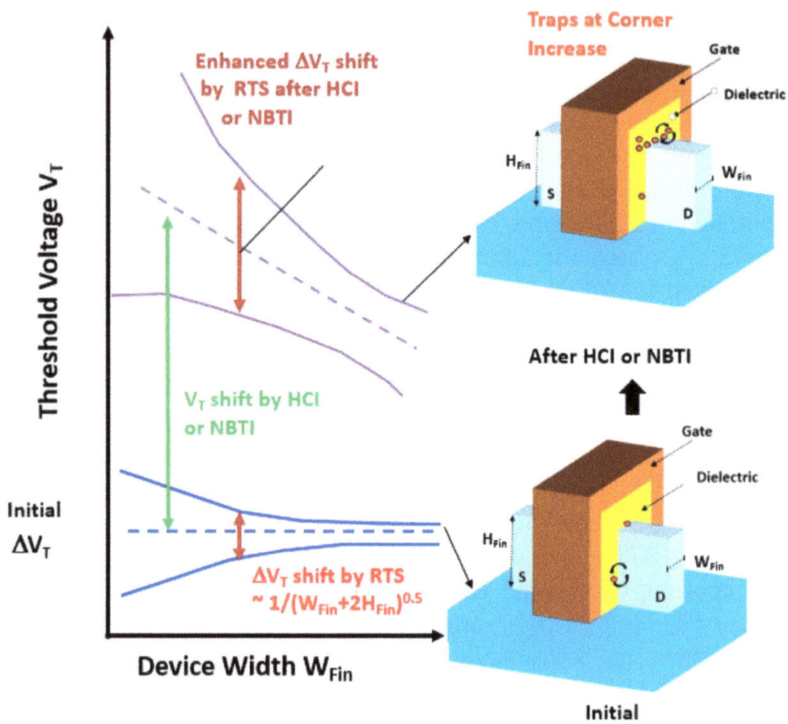

Figure 5.15. Summary of W dependence of RTS, including HC and NBTI. (After Ota *et al* [100].)

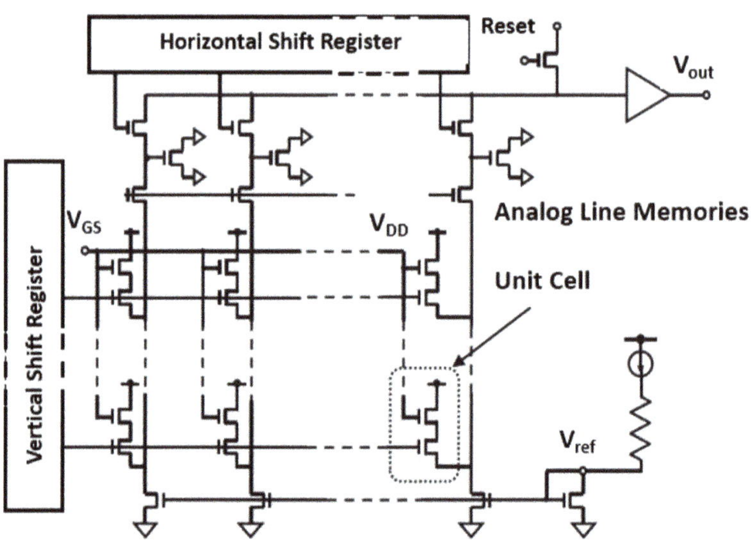

Figure 5.16. Schematic of circuit of nMOSFET source follower array test element group (TEG). (Reproduced with permission from Abe *et al* [103]. Copyright 2009 The Japan Society of Applied Physics.)

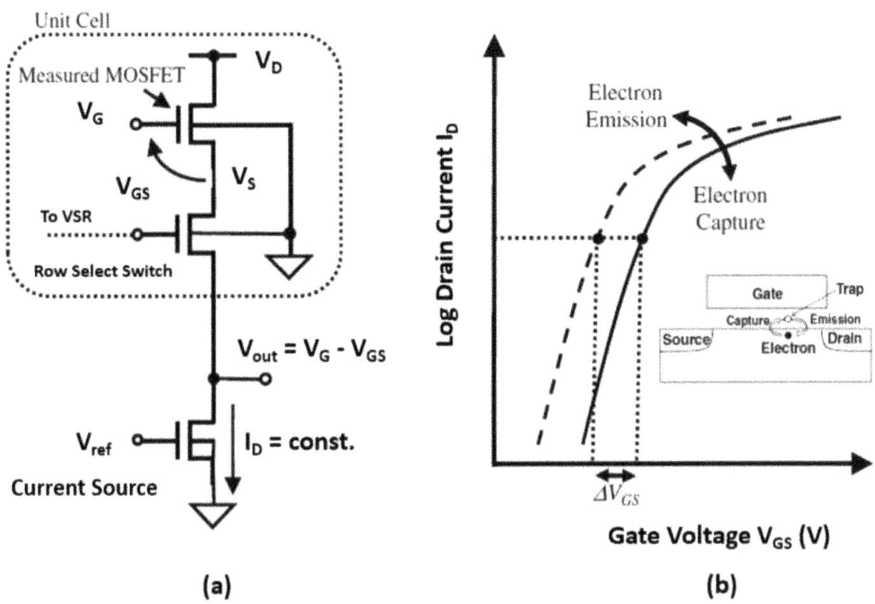

Figure 5.17. (*a*) Bias conditions and output voltage relationships of the unit cell TEG and (*b*) conceptual illustration of I_D versus V_{GS} characteristic affected by RTS. When the trap captures an electron, the threshold voltage shows a higher state compared to the case of the empty trap. (Reproduced with permission from Abe *et al* [103]. Copyright 2009 The Japan Society of Applied Physics.)

$$A_{\mathrm{m}}^2 = \frac{q^2 kTN_{\mathrm{ot}}}{C_{\mathrm{ox}}^2}\ln\left(\frac{1}{f_{\min} 2\pi t_{\mathrm{rise}}}\right) \tag{5.8}$$

with N_{ot} the area oxide-trap density (in cm^{-2}) and $f_{\min}$ and $f_{\max} = 1/(2\pi t_{\mathrm{rise}})$ the minimum and the maximum frequency for integration of the PSD of the $1/f$ fluctuations (noise) in the MOSFET.

As shown in figure 5.18, a good agreement between calculated and extracted parameters has been obtained, with a significantly higher value for p- versus nMOSFETs. This indicates a higher oxide-trap density responsible for the RTS-induced V_T shift in pMOSFETs, which is in line with the trap densities extracted from the $1/f$ noise PSD [113]. An opposite conclusion was reached in [115], where it is argued that $1/f$ noise (and RTN-induced dynamic variability) should not follow a $1/\sqrt{WL}$ dependence. This is claimed to be a natural consequence of the lognormal distribution of RTS-based LFN. Clearly, more work is needed to finally settle this argument.

Similar circuit-based approaches have been developed for addressing the dynamic variability associated with BTI and other device aging mechanisms [116–118]. One important aspect which is often overlooked is the contribution of the mobility shift induced by a charged RTS on the device current ΔI_D [96]. Usually, this is translated into an equivalent ΔV_T change. Neglecting this $\Delta\mu$ contribution can result in erroneous mean and dispersion of the V_T shifts governing stochastic NBTI. The

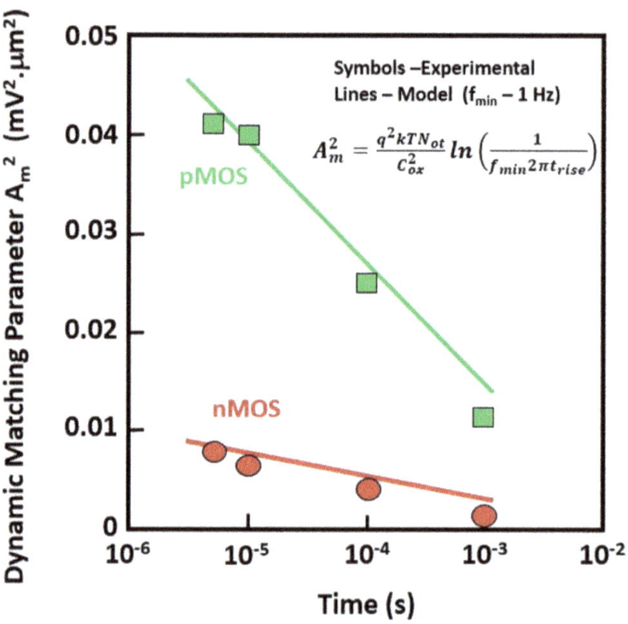

Figure 5.18. Dynamic matching parameter A_m for the median value of $\sigma(V_T)$ versus rise time experimental (symbols) versus model of equation (5.8). f_{min} is taken as ~1 Hz. (After Ioannidis *et al* [113].)

amount of error depends on the sense voltage during the stress relaxation phase [96]. Finally, it has been proposed to employ the RTS time constants of pre-existing traps, which are fairly immune to stress, as a new-concept physical unclonable function for identifying chips and/or devices [119].

5.5 Device and circuit simulation of dynamic variability

Alternatively, more insight in RTS-induced dynamic variability can be obtained from physics-based numerical simulations. This is possible on the device level to get some idea on the distributions of the specific operational parameters (V_T, etc), which then can be implemented in circuit simulators. Initial modeling works have investigated the Berkeley Short-channel IGFET Model and Simulation Program with Integrated Circuit Emphasis (SPICE) oriented modeling of RTS and related $1/f$ noise for the purpose of device and circuit simulation [120–129]. A recent example of a physics-based, stand-alone RTS simulator has been developed by the University of Texas Group at Arlington (RTSSIM) [130, 131]. It was established on experimentally identified oxide traps in the Si/SiO_2 MOS system. While this is useful in its own right, it is felt that in many cases the physical RTS models initially used are too dedicated and limited to have a broader applicability than intended.

Meanwhile, RTS-based models for BTI are being implemented in circuit simulators, including the distribution of the parameters ((kinetic) MC approach) [132–135], providing realistic predictions of the circuit performance degradation at high positive or negative gate bias. As an example, Subirats *et al* [134] have simulated the effect of a single oxide charge on the V_T, considering a bilayer gate stack (SiO_2/high-κ), combining 3D electrostatic simulations and the MC method to place traps randomly in the bilayer. They show that the 3D simulations are well reproduced by a simple 1D charge sheet model and are given by [134]

$$\Delta V_T = q\left(\frac{t_{HK}}{\varepsilon_0 \varepsilon_{HK}} + \frac{t_{IL} - x_T}{\varepsilon_0 \varepsilon_{IL}}\right) 0 \leqslant x_T \leqslant t_{IL} \tag{5.9a}$$

$$\Delta V_T = q\frac{t_{HK}}{\varepsilon_0 \varepsilon_{HK}}\left(1 - \frac{x_T - t_{IL}}{\varepsilon_0 \varepsilon_{IL}}\right) 0 \leqslant x_T \leqslant t_{IL} \tag{5.9b}$$

with t_{IL} and t_{HK} the thickness of the interfacial SiO_2 layer and the high-κ layer, respectively. From equations (5.9) it follows that the probability distribution of ΔV_T should exhibit two slopes, corresponding to traps in the different layers [134]. During NBTI stress, trapping happens mainly in the IL while for PBTI charging in the high-κ dielectric dominates. This explains why, in practice, a single distribution (slope) reproduces the experiments. However, the 'single-defect distribution' model is only valid if there is an equal number of traps in the two layers. It fails to explain a large asymmetry in the trap distribution between the two layers [134]. Therefore, under PBTI, the defect centric model still fits the bilayer case but the parameters (slopes) lose their physical meaning and become, effectively, some average of the value in each layer [134].

A general simulation framework for describing the statistical aspects of charge trapping in MOSFETs in the presence of random discrete dopants has been proposed by Bukhori *et al* [136]. The goal of the framework is to implement the effect of a statistical charge trapping distribution (as in RTS) and to study the variation of the trapped charge distribution with stressing time during NBTI, PBTI and HC degradation. A 3D atomistic numerical simulation of large statistical samples of realistic devices has been used to investigate the current changes and the evolution of the statistical distribution of MOSFET parameters subject to a progressive degradation and the corresponding increase in the stochastic numbers of trapped charge [136]. The calculation assumes acceptor type traps in Si/SiO_2 and its impact on ΔV_T and $\Delta I_D/I_D$. The simulation of progressive degradation has been emulated by increasing the average density of trapped charge in the Poisson distribution. An almost linear increase with average trapped charge has been found for the ΔV_T in figure 5.19, while the distribution widens at the same time [136]. It was shown that ignoring RDF does not produce realistic V_T distributions with increasing trap density, i.e. they do not agree with experimental observations. Therefore, it is the interplay between RDF and the statistical distribution of traps which explains the experimentally observed trends [136].

Moreover, it has been shown that RDF enhances the maximum ΔV_T by a factor of 2 compared with a continuous doping model. A slightly negative correlation between the initial V_T and ΔV_T with stress follows from the calculations. This means that the V_T variations due to RDF and trapping do not necessarily add up. The simulations have also been compared to a simple analytical model for the standard deviation of the V_T distribution, given by [136]

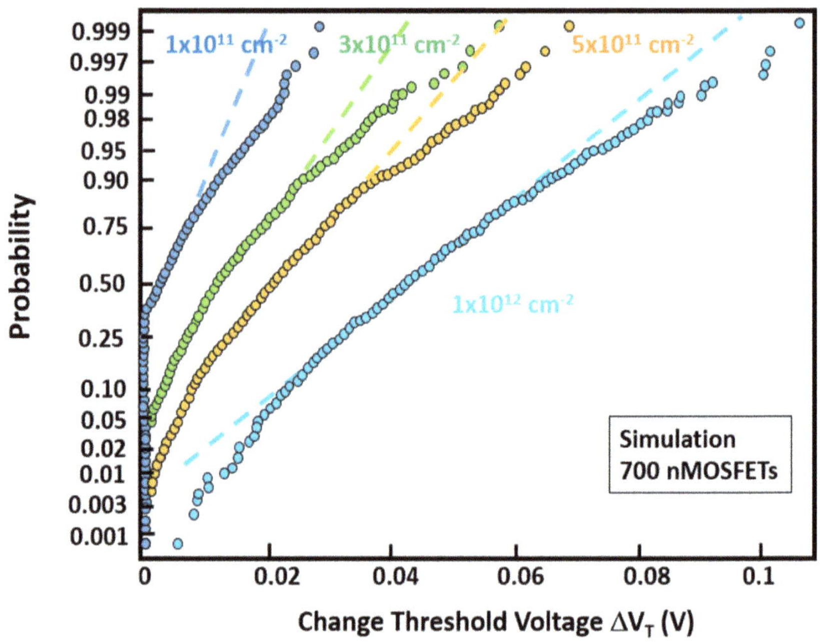

Figure 5.19. Distribution of threshold voltage change in 700 MOSFETs subjected to progressive increase of the sheet density of trapped electrons. (After Bukhori *et al* [136].)

$$\sigma \Delta V_{\mathrm{T}} = \frac{q t_{\mathrm{ox}} \sqrt{N_{\mathrm{ot}}}}{\varepsilon_0 \varepsilon_{\mathrm{ox}} \sqrt{WL}} \tag{5.10}$$

with N_{ot} the density of trapped charges. As can be seen in figure 5.20, the simple model of equation (5.10) underestimates the calculated average ΔV_{T} as it does not consider the random position of the traps with respect to the random dopant arrangement, introducing significant variation in ΔV_{T} [136].

Another aspect which becomes obvious from kinetic MC-based device simulations of oxide-trap-induced V_{T} shifts is that trap interactions cannot be neglected [68, 137], as can be observed in figures 5.21 and 5.22. From figure 5.21 one derives that multiple charge trapping effects (here eight oxide traps are considered) shift the capture time constant to longer times [137]. The magnitude of the effect depends on the local position of the traps. This implies that the trap dynamics are heavily affected by other charged defects and cannot be considered independent. On the other hand, from figure 5.22 it follows that the ΔV_{T} distributions are smooth and narrow, showing fewer mutual interactions [133]. The maximum impact of each trap on ΔV_{T} depends on the overall configuration of filled and empty traps. This yields either an increase or a decrease of the ΔV_{T} obtained for independent traps. This difference in response of the RTS (BTI) time constants and amplitude is related to the fact that the capture time constant is correlated with the local charge density underneath the trap, while ΔV_{T} is correlated to the non-local presence of a percolative conduction path connecting the source and drain [137]. Another

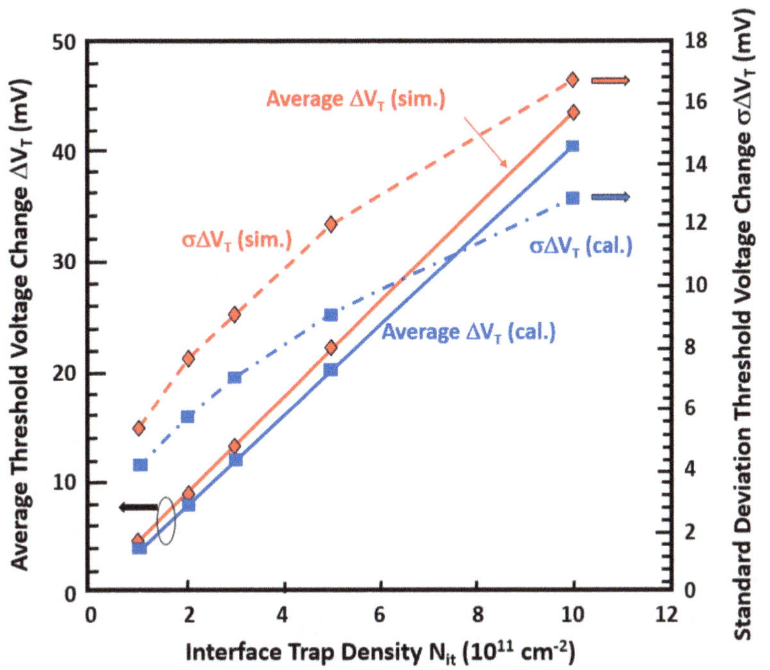

Figure 5.20. Average threshold voltage change and its standard deviation as a function of interface trap density obtained from the simulations compared against calculations. (After Bukhori *et al* [136].)

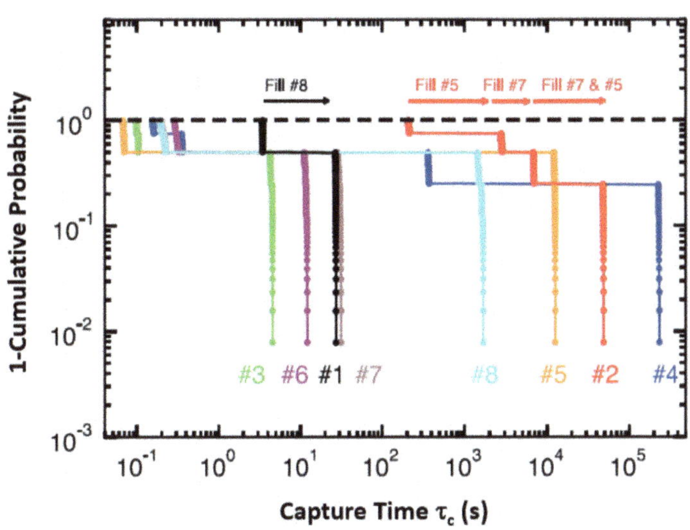

Figure 5.21. Cumulative probability distribution of capture times for one device with eight traps at $V_{GS} = V_T$, illustrating their sensitivity on the occupancy status of other traps. (Reproduced with permission from Amoroso *et al* [137]. Copyright 2013 Elsevier.)

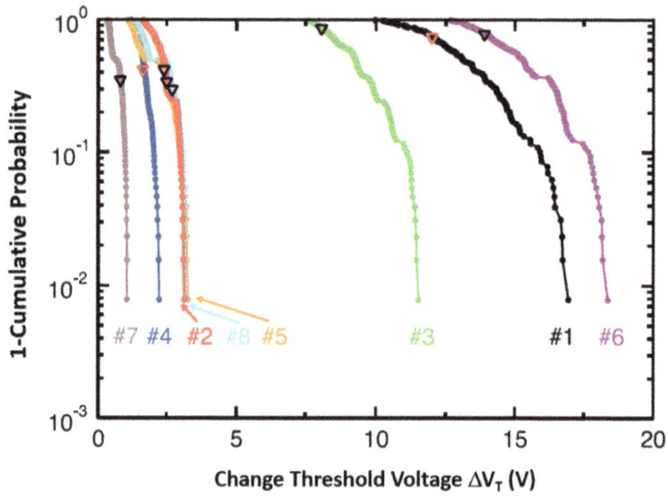

Figure 5.22. Cumulative probability distribution function of single-charge trap-induced ΔV_T for the same device as figure 5.19, illustrating the sensitivity of each single trap on the occupancy status of other traps. Triangles indicate the ΔV_T given by each trap when the other traps are not active. (Reproduced with permission from Amoroso *et al* [137]. Copyright 2013 Elsevier.)

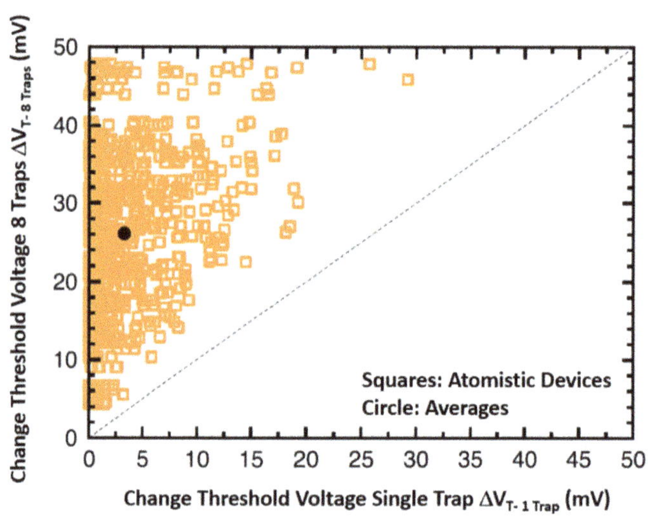

Figure 5.23. ΔV_T due to eight traps (BTI) as a function of ΔV_T due to a single (RTN) trap for 100 atomistic devices. The black symbol indicates the point of coordinates $(\langle \Delta V_{T\text{-}1}, \Delta V_{T\text{-}8} \rangle)$. Note that $\langle \Delta V_{T\text{-}8} \rangle = 8 \times \langle \Delta V_{T\text{-}1} \rangle$. Note also that all the points are scattered above the graph diagonal because $\Delta V_{T\text{-}8} \geqslant \Delta V_{T\text{-}1}$. (Reproduced with permission from Amoroso *et al* [137]. Copyright 2013 Elsevier.)

important conclusion is that there is no correlation between RTS and BTI (figure 5.23) [137]: the ΔV_T due to a single trap (RTN) is not correlated to the ΔV_T due to many traps (BTI). This is related to the percolative nature of the source-to-drain conduction in nanoscale transistors. Therefore, devices strongly affected by RTN are not necessarily strongly affected by BTI and vice versa.

A further step is to develop methods for variability-aware design of circuits, either relying on standard programs such as SPICE, with dedicated RTS-based device models [138–145], or relying on a dedicated simulation tool, such as MUSTARD [146, 147]. A comprehensive review of RTN/BTI modeling from device to circuit level has recently been published by the Glasgow modeling group [148]. Finally, the single-trap effects on the stability of heterojunction TFET SRAMs have been modeled in [149].

References

[1] Dierickx B and Simoen E 1992 The decrease of 'random telegraph signal' noise in metal–oxide–semiconductor field-effect transistors when cycled from inversion to accumulation *J. Appl. Phys.* **71** 2028–9

[2] van der Wel A P, Klumperink E A M and Nauta B 2001 Effect of switched biasing on 1/*f* noise and random telegraph signals in deep-submicron MOSFETs *Electron. Lett.* **37** 55–6

[3] van der Wel A P, Klumperink E A M, Vandamme L K J and Nauta B 2003 Modeling random telegraph noise under switched bias conditions using cyclostationary RTS noise *IEEE Trans. Electron Devices* **50** 1378–84

[4] Kolhatkar J S, Hoekstra E, Salm C, van der Wel A P, Klumperink E A M, Schmitz J and Wallinga H 2004 Modeling of RTS noise in MOSFETs under steady-state and large-signal excitation *Tech. Dig. Int. Electron Devices Meeting—IEDM04* (New York: IEEE) pp 759–62

[5] van der Wel A P, Klumperink E A M, Hoekstra E and Nauta B 2005 Relating random telegraph signal noise in metal–oxide–semiconductor transistors to interface trap energy distribution *Appl. Phys. Lett.* **87** 183507

[6] Roy A S and Enz C C 2004 Analytically modelling of large signal cyclo-stationary low frequency noise for arbitrary periodic input *Proc. ESSDERC-ESSCIRC 2004* (New York: IEEE), pp 455–8

[7] Wirth G, da Silva R, Srinivasan P, Krick J and Brederlow R 2009 Statistical model for MOSFET low-frequency noise under cyclo-stationary conditions *Tech Dig. Int. Electron Devices Meeting—IEDM09* (New York: IEEE), pp 715–8

[8] Miki H, Tega N, Ren Z, D'Emic C P, Zhu Y, Frank D J, Guillorn M A, Park D-G, Haensch W and Torii K 2010 Hysteretic drain-current behavior due to random telegraph noise in scaled-down FETs with high-κ/metal-gate stacks *Tech. Dig. Int. Electron Devices Meeting—IEDM10* (New York: IEEE), pp 620–3

[9] Grasser T, Reisinger H, Wagner P-J, Schanovsky F, Goes W and Kaczer B 2010 The time dependent defect spectroscopy (TDDS) for the characterization of the bias temperature instability *Proc. of the Annual International Reliability Physics—IRPS10* (New York: IEEE) pp 16–25

[10] Grasser T, Reisinger H, Wagner P-J and Kaczer B 2010 Time-dependent defect spectroscopy for characterization of border traps in metal–oxide–semiconductor transistors *Phys. Rev.* B **82** 245318

[11] Zanolla N, Šiprak D, Tiebout M, Baumgartner P, Sangiorgi E and Fiegna C 2010 Reduction of RTS noise in small-area MOSFETs under switched bias conditions and forward substrate bias *IEEE Trans. Electron Devices* **57** 1119–28

[12] Guo S *et al* 2014 DTMOS mode as an effective solution of RTN suppression for robust device/circuit co-design *Tech. Dig. Int. Electron Devices Meeting—IEDM14* (New York: IEEE) pp 319–22

[13] Feng W, Yamada K and Ohmori K 2013 Random telegraph noise induced drain-current fluctuation during dynamic gate bias in Si MOSFETs *Proc. Int. Conf. on 1/f Noise and Fluctuations—ICNF 2013* (New York: IEEE) pp 1–4

[14] Feng W, Dou C M, Niwa M, Yamada K and Ohmori K 2014 Impact of random telegraph noise profiles on drain-current fluctuation during dynamic gate bias *IEEE Electron Device Lett.* **35** 3–5

[15] Zou J *et al* 2013 Deep understanding of AC RTN in MuGFETs through new characterization method and impacts on logic circuits *Symp. on VLSI Technol. Dig. Techn. Papers* (New York: IEEE) pp T186–7

[16] Ren P *et al* 2013 New observations on complex RTN in scaled high-κ/metal-gate MOSFETs—the role of defect coupling under DC/AC condition *Tech. Dig. Int. Electron Devices Meeting—IEDM13* (New York: IEEE) pp 778–81

[17] Zou J *et al* 2012 New insights into AC RTN in scaled high-κ/metal-gate MOSFETs under digital circuit operations *Symp. on VLSI Technol. Dig. Techn. Papers* (New York: IEEE) pp 139–40

[18] Matsumoto T, Kobayashi K and Onodera H 2012 Impact of random telegraph noise on CMOS logic delay uncertainty under low voltage operation *Tech. Dig. Int. Electron Devices Meeting—IEDM12* (New York: IEEE) pp 581–4

[19] Matsumoto T, Kobayashi K and Onodera H 2013 Impact of body-biasing technique on RTN-induced delay fluctuation *Jpn. J. Appl. Phys.* **52** 04CE05

[20] Tang Q, Wang X, Keane J and Kim C H 2013 RTN induced frequency shift measurement using a ring oscillator based circuit *Symp. on VLSI Technol. Dig. Tech. Papers* (New York: IEEE) pp T188–9

[21] Prati E, Fanciulli M, Calderoni A, Ferrari G and Sampietro M 2008 Effect of microwave irradiation on the emission and capture dynamics in silicon metal oxide semiconductor field effect transistors *J. Appl. Phys.* **103** 104502

[22] Xiao M, Martin I and Jiang H W 2003 Probing the spin state of a single electron trap by random telegraph signal *Phys. Rev. Lett.* **91** 078301

[23] Ferrari G, Fumagalli L, Sampietro M, Prati E and Fanciulli M 2005 dc modulation in field-effect transistors operating under microwave irradiation for quantum readout *J. Appl. Phys.* **98** 044505

[24] Yonamoto Y and Akamatsu N 2011 Direct observation in stressed SiO_2 with electrically detected magnetic resonance *Appl. Phys. Lett.* **98** 232903

[25] Baron F A, Zhang Y and Wang K L 2003 Observation of magnetic-field-enhanced source current of accumulated p-channel metal–oxide–semiconductor field-effect transistors *Appl. Phys. Lett.* **82** 3547–9

[26] Avellan A, Schroeder D and Krautschneider W 2003 Modeling random telegraph signals in the gate current of metal–oxide–semiconductor field-effect transistors after oxide breakdown *J. Appl. Phys.* **94** 703–8

[27] Farmer K R and Buhrman R A 1989 Defect dynamics and wear-out in thin silicon oxides *Semicond. Sci. Technol.* **4** 1084–108

[28] Simoen E and Claeys C 1999 Reliability aspects of the low-frequency noise behaviour of submicron CMOS technologies *Semicond. Sci. Technol.* **14** R61–R71

[29] Omura Y and Komiya K 2002 Transport characteristics of posthard breakdown thin silicon oxide films and consideration of physical models *J. Appl. Phys.* **91** 4298–306

[30] Li X, Tung C H, Pey K L and Lo V L 2009 The physical origin of random telegraph noise after dielectric breakdown *Appl. Phys. Lett.* **94** 132904

[31] Armand J, Martinez F, Benoit P, Valenza M, Vincent E, Huard V and Rocherau K 2009 dc and low frequency noise analysis of Fowler–Nordheim stress of n-channel metal–oxide–semiconductor field-effect transistors processed in a 65 nm technology *J. Vac. Sci. Technol.* B **27** 1129–31

[32] Li X, Tung C H and Pey K L 2008 The nature of dielectric breakdown *Appl. Phys. Lett.* **93** 072903

[33] Li X, Tung C H, Pey K L and Lo V L 2009 The physical origin of random telegraph noise after dielectric breakdown *Appl. Phys. Lett.* **94** 132904

[34] Padovani A, Morassi L, Raghavan N, Larcher L, Liu W, Pey K L and Bersurker G 2010 A physical model for post-breakdown digital gate current noise *IEEE Electron Device Lett.* **31** 1032–4

[35] Brière O, Chroboczek J A and Ghibaudo G 1996 Random telegraph signal in the quasi-breakdown current of MOS capacitors *Proc. of ESSDERC 1996* (New York: IEEE) pp 759–62

[36] Vandewalle N, Ausloos M, Houssa M, Mertens P W and Heyns M M 1999 Non-Gaussian behavior and anticorrelations in ultrathin gate oxides after soft breakdown *Appl. Phys. Lett.* **74** 1579–81

[37] Houssa M, Vandewalle N, Nigam T, Ausloos M, Mertens P W and Heyns M M 1998 Analysis of the gate voltage fluctuations in ultra-thin gate oxides after soft breakdown *Tech. Dig. Int. Electron Devices Meeting—IEDM98* (New York: IEEE) pp 909–12

[38] Chung S S and Chang C M 2008 The investigation of capture/emission mechanism in high-κ gate dielectric soft breakdown by gate current random telegraph noise approach *Appl. Phys. Lett.* **93** 213502

[39] Choi H-S, Hong S-H, Baek R-H, Lee K-T, Kang C-Y, Jammy R, Lee B-H, Jung S-W and Jeong Y-H 2009 Low-frequency noise after channel soft oxide breakdown in HfLaSiO gate dielectric *IEEE Electron Device Lett.* **30** 523–5

[40] Liu W, Padovani A, Larcher L, Raghavan N and Pey K L 2014 Analysis of correlated gate and drain random telegraph noise in post-soft breakdown TiN/HfLaO/SiO_x nMOSFETs *IEEE Electron Device Lett.* **35** 157–9

[41] Kim H, Oh B, Kim K, Cha S-Y, Jeong J-G, Hong S-J, Lee J-H, Park B-G and Shin H 2010 Random telegraph signal-like fluctuation created by Fowler–Nordheim stress in gate-induced drain leakage current of the saddle type dynamic random access memory cell transistor *Jpn. J. Appl. Phys.* **49** 094102

[42] Raghavan N, Leong K, Liu W and Bosman M 2010 Postbreakdown gate-current low-frequency noise spectrum as a detection tool for high-κ and interfacial layer breakdown *IEEE Electron Device Lett.* **31** 1035–7

[43] Liu W H, Pey K L, Raghavan N, Wu X, Bosman M and Kauerauf T 2011 Random telegraph noise reduction in metal gate high-κ gate stacks by bipolar switching and the performance boosting technique *Proc. Annual International Reliability Physics Symp.—IRPS11* (New York: IEEE) pp 182–9

[44] Liu W H, Pey K L, Raghavan N, Wu X and Bosman M 2012 Triggering voltage for post-breakdown random telegraph noise in HfLaO dielectric metal gate metal–oxide–semiconductor field effect transistors and its reliability implications *J. Appl. Phys.* **111** 0241011

[45] Lo C C, Persaud A, Dhuey S, Olynick D, Borondics F, Martin M C, Bechtel H A, Bokor J and Schenkel T 2009 Device fabrication and transport measurements of FinFETs built with ^{28}Si SOI wafers toward donor qubits in silicon *Semicond. Sci. Technol.* **24** 105022

[46] Raghavan N, Bosman M and Pey K L 2015 Spectroscopy of SILC trap location and spatial correlation study of percolation path in the high-κ and interfacial layer *Proc. Annual International Reliability Physics Symp.—IRPS15* (New York: IEEE) pp 5A2.1–2.7

[47] Komiya K and Omura Y 2002 Spectroscopic characterization of stress-induced leakage current in sub 5-nm-thick silicon oxide film *J. Appl. Phys.* **92** 2593–601

[48] Ishida T, Tega N, Mori Y, Miki H, Kume H, Torii K, Muraguchi M, Takada Y, Shiraishi K and Yamada R 2008 A new insight into the dynamic fluctuation mechanism of stress-induced leakage current *Proc. of the 46th Annual International Reliability Physics Symp.—IRPS08* (New York: IEEE) pp 604–9

[49] Teramoto A, Kumagai Y, Abe K, Fujisawa T, Watabe S, Suwa T, Miyamoto N, Sugawa S and Ohmi T 2009 Stress-induced leakage current and random telegraph signal *J. Vac. Sci. Technol.* B **27** 435–8

[50] Ishida T, Tega N, Mori Y, Miki H, Mine T, Kume H, Torii K, Yamada R and Shiraishi K 2014 Mechanism of state transition of a defect causing random-telegraph-noise-induced fluctuation in stress-induced leakage current of SiO_2 films *Jpn. J. Appl. Phys.* **53** 08LB01

[51] Teramoto A, Inatsula T, Obara T, Akagawa N, Kuroda R, Sugawa S and Ohmi T 2014 Demonstrating individual leakage path from random telegraph signal of stress induced leakage current *Proc. Annual International Reliability Physics Symp.—IRPS14* (New York: IEEE) pp GD1.1–1.5

[52] Hsieh E R and Chung S S 2015 The understanding on the evolution of stress-induced gate leakage in high-κ dielectric metal–oxide-field-effect transistor by random-telegraph-noise measurement *Appl. Phys. Lett.* **107** 243506

[53] Kwon H-M, Bok J-D, Han I-S, Park S-U, Jung Y J, Jang J-H, Ko S-Y, Lee W-M, Lee G-W and Lee H-D 2011 Dependence of hot carrier reliability and low frequency noise on channel stress in nanoscale n-channel metal–oxide–semiconductor field-effect transistors *Jpn. J. Appl. Phys.* **50** 10B0110B01

[54] Fang P, Hung K K, Ko P K and Hu C 1990 Characterizing a single hot-electron-induced trap in submicron MOSFET using random telegraph signal *Symp. VLSI Technol. Dig. Techn. Papers* (New York: IEEE) pp 37–8

[55] Fang P, Hung K K, Ko P K and Hu C 1991 Hot-electron-induced traps studied through the random telegraph noise *IEEE Electron Device Lett.* **12** 273–5

[56] Muto H, Tsai M-H and Ma T-P 1991 Random telegraph signals in small MOSFET's after x-ray irradiation *IEEE Trans. Nucl. Sci.* **38** 1116–23

[57] Simoen E, Dierickx B and Claeys C 1993 Hot-carrier degradation of the random telegraph signal amplitude in submicrometer Si MOSTs *Appl. Phys.* A **57** 283–9

[58] Simoen E and Claeys C 1997 Hot-carrier stress effects on the amplitude of random telegraph signals in small area Si p-MOSFETs *Microelectron. Reliab.* **37** 1015–9

[59] Cho H-J, Son Y, Oh B-C, Lee S, Lee J-H, Park B-G and Shin H 2010 Study on time constants of random telegraph noise in gate leakage current through hot-carrier stress test *IEEE Electron Device Lett.* **31** 1029–31

[60] Liu C, Lee K T, Pae S and Park J 2014 New observations on hot carrier induced dynamic variation in nano-scaled SiON/poly, HK/MG and FinFET devices based on on-the-fly HCI

technique: the role of single trap induced degradation *Tech. Dig. Int. Electron Devices Meeting—IEDM 14* (New York: IEEE) pp 836–9
[61] Ota K, Saitoh M, Tanaka C, Matsushita D and Numata T 2014 Systematic study of RTN in nanowire transistor and enhanced RTN by hot carrier injection and negative bias temperature instability *Symp. on VLSI Technol. Dig. Techn. Papers* (New York: IEEE) pp 1–2
[62] Kamei M, Takao Y, Eriguchi K and Ono K 2014 Effects of plasma-induced charging damage on random telegraph noise in metal–oxide–semiconductor field-effect transistors with SiO_2 and high-κ gate dielectrics *Jpn. J. Appl. Phys.* **53** 03DF02
[63] Grasser T *et al* 2011 The paradigm shift in understanding the bias temperature instability: from reaction-diffusion to switching oxide traps *IEEE Trans. Electron Devices* **58** 3652–66
[64] Kaczer B, Grasser T, Martin-Martinez J, Simoen E, Aoulaiche M, Roussel Ph J and Groeseneken G 2009 NBTI from the perspective of defect states with widely distributed time scales *Proc. 47th Annual International Reliability Physics Symp.—IRPS09* (New York: IEEE) pp 55–60
[65] Kaczer B, Grasser T, Franco J, Toledano-Luque M, Roussel Ph J, Cho M, Simoen E and Groeseneken G 2011 Recent trends in bias temperature instability *J. Vac. Sci. Technol.* B **29** 01AB01
[66] Kirton M J, Uren M J, Collins S, Schulz M, Karmann A and Scheffer K 1989 Individual defects at the Si:SiO_2 interface *Semicond. Sci. Technol.* **4** 1116–26
[67] Karwath A and Schulz M 1988 Deep level transient spectroscopy on single, isolated interface traps in field-effect transistors *Appl. Phys. Lett.* **52** 634–6
[68] Grasser T 2012 Stochastic charge trapping in oxides: from random telegraph noise to bias temperature instabilities *Microelectron. Reliab.* **52** 39–70
[69] Grasser T, Rott K, Reisinger H, Wagner P-J, Goes W, Schanovsky F, Waltl M, Toledano-Luque M and Kaczer B 2013 Advanced characterization of oxide traps: the dynamic time-dependent defect spectroscopy *Proc. Annual International Reliability Physics Symp.—IRPS13* (New York: IEEE) pp 2D2.1–7
[70] Reisinger H, Grasser T, Gustin W and Schlünder C 2010 The statistical analysis of individual defects constituting NBTI and its implications for modeling DC- and AC-stress *Proc. Annual International Reliability Physics Symp.—IRPS10* (New York: IEEE) pp 7–15
[71] Amarasinghe N V and Çelik-Butler Z 2000 Complex random telegraph signals in 0.06 μm^2 MDD n-MOSFETs *Solid-State Electron.* **44** 1013–9
[72] Toledano-Luque M, Kaczer B, Roussel Ph J, Franco J, Ragnarsson L-Å, Grasser T and Groeseneken G 2011 Depth localization of positive charge trapped in silicon oxynitride field effect transistors after positive and negative gate bias temperature stress *Appl. Phys. Lett.* **98** 183506
[73] Grasser T, Reisinger H, Goes W, Aichinger Th, Hehenberger Ph, Wagner P-J, Nelhiebel M, Franco J and Kaczer B 2009 Switching oxide traps as the missing link between negative bias temperature instability and random telegraph noise *Tech. Dig. Int. Electron Devices Meeting—IEDM09* (New York: IEEE) pp 729–32
[74] Grasser T *et al* 2010 Recent advances in understanding the bias temperature instability *Tech. Dig. Int. Electron Devices Meeting—IEDM10* (New York: IEEE) pp 82–5
[75] Grasser T, Rott K, Reisinger H, Waltl M, Franco J and Kaczer B 2014 A unified perspective of RTN and BTI *Proc. Annual International Reliability Physics Conf.—IRPS14* (New York: IEEE) pp 4A5.1–5

[76] Velamala J B, Sutaria K B, Sato T and Cao Y 2012 Aging statistics based on trapping/detrapping: silicon evidence, modeling and long-term prediction *Proc. Annual International Reliability Physics Symp.—IRPS12* (New York: IEEE) pp 2F.2.1–2F2.5

[77] Grasser T, Rott K, Reisinger H, Waltl M, Schanovsky F and Kaczer B 2014 NBTI in nanoscale MOSFETs—the ultimate modeling benchmark *IEEE Trans. Electron Devices* **61** 3586–93

[78] Chiu J-P, Li C-W and Wang T 2012 Characterization and modeling of trap number and creation time distributions under negative-bias-temperature stress *Appl. Phys. Lett.* **101** 082906

[79] Amoroso S M, Gerrer L, Markov S, Adamu-Lema F and Asenov A 2012 Comprehensive statistical comparison of RTN and BTI in deeply scaled MOSFETs by means of 3D 'atomistic' simulation *Proc. ESSDERC '12* (New York: IEEE) pp 109–12

[80] Grasser T, Wagner P-J, Reisinger H, Aichinger Th, Pobegen G, Nelhiebel M and Kaczer B 2011 Analytic modeling of the bias temperature instability using capture/emission time maps *Tech. Dig. Int. Electron Devices Meeting IEDM 11* (New York: IEEE) pp 618–21

[81] Grasser T, Kaczer B, Reisinger H, Wagner P-J and Toledano-Luque M 2012 On the frequency dependence of the bias temperature instability *Proc. Annual International Reliability Physics Symp.—IRPS12* (New York: IEEE) pp XT.8.1–XT.8.7

[82] Duan M, Zhang J F, Ji Z and Zhang W 2016 Defects for random telegraph noise and negative bias temperature instability *Proc. China Semiconductor Technology Int. Conf.—CSTIC 2016* (New York: IEEE) pp 1–4

[83] Toledano-Luque M, Kaczer B, Roussel Ph J, Grasser T, Wirth G I, Franco J, Vrancken C, Horiguchi N and Groeseneken G 2011 Response of a single trap to AC negative bias temperature stress *Proc. Annual International Reliability Physics Symp.—IRPS11* (New York: IEEE) pp 364–71

[84] Zhao K, Stathis J H, Linder B P, Cartier E and Kerber A 2011 PBTI under dynamic stress: from a single defect point of view *Proc. Annual International Reliability Physics Symp.—IRPS11* (New York: IEEE) pp 372–80

[85] Huang D C and Gong J 2012 Investigation of P/NBTI in high-κ gate dielectric MOSFETs by I_d-RTS *Extended Abstracts of the 2012 Int. Conf. on Solid State Devices and Materials* (Kyoto, Japan) pp 126–7

[86] Chiu J-P, Liu Y-H, Hsieh H-D, Li C-W, Chen M-C and Wang T 2013 Statistical characterization and modeling of the temporal evolutions of ΔV_t distribution in NBTI recovery in nanometer MOSFETs *IEEE Trans. Electron Devices* **60** 978–84

[87] Huang D-C, Gong J and Huang C-F 2013 Investigation of positive and negative bias temperature instability of high-κ dielectric metal gate metal–oxide–semiconductor-field-effect-transistors by random telegraph signal *Jpn. J. Appl. Phys.* **52** 036501

[88] Tega N, Miki H, Mine T, Ohmori K and Yamada K 2014 Statistical analysis of relationship between negative-bias temperature instability and random telegraph noise in small p-channel metal–oxide–semiconductor field-effect transistors *Jpn. J. Appl. Phys.* **53** 034202/–5

[89] Franco J, Kaczer B, Mitard J, Toledano-Luque M, Roussel Ph J, Witters L, Grasser T and Groeseneken G 2013 NBTI reliability of SiGe and Ge channel pMOSFETs with SiO_2/HfO_2 dielectric stack *IEEE Trans. Device and Mater. Reliab.* **13** 497–506

[90] Zhang L, Liu C, Wang R, Huang R, Yu T, Zhuge J, Kirsch P, Tseng H-H and Wang Y 2010 Characteristics of gate current random telegraph signal noise in $SiON/HfO_2/TaN$ p-type metal–oxide–semiconductor field-effect transistors under negative bias temperature instability stress conditions *Jpn. J. Appl. Phys.* **49** 04DC08

[91] Chen C-Y, Ran Q, Cho H-J, Kerber A, Liu Y, Lin M-R and Dutton R W 2011 Correlation of I_d- and I_g-random telegraph noise to positive bias temperature instability in scaled high-κ/metal gate n-type MOSFETs *Proc. Annual International Reliability Physics Symp.—IRPS11* (New York: IEEE) pp 190–5

[92] Han I-S *et al* 2011 Effect of nitrogen concentration on low frequency noise and negative bias temperature instability of p-channel metal–oxide–semiconductor field-effect transistors with nitride gate oxide *Jpn. J. Appl. Phys.* **50** 10OB0310OB03

[93] Liu C, Lee K T, Lee H, Kim Y, Pae S and Park J 2014 New observations on the random telegraph noise induced Vth variation in nano-scale MOSFETs *Proc. Annual International Reliability Physics Symp.—IRPS14* (New York: IEEE) pp XT.17.1–17.5

[94] Waltl M, Goes W, Rott K, Reisinger H and Grasser T 2014 A single-trap study of PBTI in SiON nMOS transistors: similarities and differences to the NBTI/pMOS case *Proc. Annual International Reliability Physics Symp.—IRPS14* (New York: IEEE) pp XT.18.1–18.5

[95] Wu P C, Hsieh E R, Lu P Y, Chung S S, Chang K Y, Liu C H, Ke J C, Yang C W and Tsai C T 2014 The observations of BTI-induced RTN traps in inversion and accumulation modes on HfO_2 high-κ metal gate 28 nm CMOS devices *Int. Symp. on VLSI Technol. Dig. Techn. Papers* (New York: IEEE) pp 1–2

[96] Chaudhary A, Kaczer B, Roussel P J, Chiarella T, Horiguchi N and Mahapatra S 2015 Time dependent variability in RMG-HKMG FinFETs: impact of extraction scheme on stochastic NBTI *Proc. Annual International Reliability Physics Symp.—IRPS15* (New York: IEEE) pp 3B.4.1–4.8

[97] Chen J, Higashi Y, Kato K and Mitani Y 2014 Further understanding on random telegraph noise through comprehensive studies on large time constant variation and its strong correlations to thermal activation energies *Symp. on VLSI Technol. Dig. Techn. Papers* (New York: IEEE) pp 1–2

[98] Franco J *et al* 2014 RTN and PBTI-induced time-dependent variability of replacement metal-gate high-κ InGaAs FinFETs *Tech. Dig. Int. Electron Devices Meeting—IEDM14* (New York: IEEE) pp 506–9

[99] Marko P, Meneghini M, Bychikhin S, Marcon D, Meneghesso G, Zanoni E and Pogany D 2012 IV, noise and electroluminescence analysis of stress-induced percolation paths in AlGaN/GaN high electron mobility transistors *Microelectron. Reliab.* **52** 2194–9

[100] Ota K, Saitoh M, Tanaka C, Matsushita D and Numata T 2015 Experimental study of random telegraph noise in trigate nanowire MOSFETs *IEEE Trans. Electron Devices* **62** 3799–804

[101] Abe K, Sugawa S, Watabe S, Miyamoto N, Teramoto A, Kamata Y, Shibusawa K, Toita M and Ohmi T 2007 Random telegraph signal statistical analysis using a very large-scale array TEG with 1 M MOSFETs *Symp. on VLSI Techn. Dig. of Techn. Papers* pp 210–1

[102] Watabe S, Sugawa S, Teramoto A and Ohmi T 2007 New statistical evaluation method for the variation of metal–oxide–semiconductor field-effect transistors *Jpn. J. Appl. Phys.* **46** 2054–7

[103] Abe K, Fujisawa T, Teramoto A, Watabe S, Sugawa S and Omi T 2009 Anomalous random telegraph signal extractions from a very large number of n-metal oxide semiconductor field-effect transistors using test element groups with 0.47 Hz–3.0 MHz sampling frequency *Jpn. J. Appl. Phys.* **48** 04C044

[104] Realov S and Shepard K L 2010 Random telegraph noise in 45-nm CMOS: analysis using an on-chip test and measurement system *Tech. Dig. Int. Electron Devices Meeting—IEDM10* (New York: IEEE) pp 624–7

[105] Monzio Compagnoni C, Gusmeroli R, Spinelli A S, Lacaita A L, Bonanomi M and Visconti A 2007 Statistical investigation of random telegraph noise I_D instabilities in flash cells at different initial trap filling conditions *Proc. 45th Annual International Reliability Physics Symp.—IRPS07* (New York: IEEE) pp 161–6

[106] Miki H, Osabe T, Tega N, Kotabe A, Kurata H, Tokami K, Ikeda Y, Kamohara S and Yamada R 2007 Quantitative analysis of random telegraph signals as fluctuations of threshold voltages in scaled flash memory cells *Proc. of the 45th Annual International Reliability Physics Symp.—IRPS07* (New York: IEEE) pp 29–35

[107] Chimenton A, Zambelli C and Olivo P 2009 A new automated methodology for random telegraph signal identification and characterization: a case study on phase change memory arrays *Proc. of the 47th Annual International Reliability Physics Symp.—IRPS09* (New York: IEEE) pp 128–33

[108] Kang D *et al* 2011 A new approach of NAND flash cell trap analysis using RTN characteristics *Symp. on VLSI Technol. Dig. of Techn. Papers* (New York: IEEE) pp 206–7

[109] Nagumo T, Takeuchi K, Yokogawa S, Imai K and Hayashi Y 2009 New analysis methods for comprehensive understanding of random telegraph noise *Tech. Dig. Int. Electron Devices Meeting—IEDM09* (New York: IEEE) pp 759–62

[110] Fujisawa T, Abe K, Watabe S, Miyamoto N, Teramoto A, Sugawa S and Ohmi T 2010 Analysis of hundreds of time constant ratios and amplitudes of random telegraph signal with very large scale array test pattern *Jpn. J. Appl. Phys.* **49** 04DC06

[111] Chimenton A, Zambelli C and Olivo P 2010 A new methodology for two-level random-telegraph-noise identification and statistical analysis *IEEE Electron Device Lett.* **31** 612–4

[112] Obara T, Yonezawa A, Teramoto A, Kuroda R, Sugawa S and Ohmi T 2014 Extraction of time constants ratio over nine orders of magnitude for understanding random telegraph noise in metal–oxide–semiconductor field-effect transistors *Jpn. J. Appl. Phys.* **53** 04EC19

[113] Ioannidis E G, Haendler S, Theodorou C G, Planes N, Dimitriadis C A and Ghibaudo G 2014 Statistical analysis of dynamic variability in 28 nm FD-SOI MOSFETs *Proc. European Solid-State Devices Conf.—ESSDERC 2014* (New York: IEEE) pp 214–7

[114] Theodorou C G, Ionanidis E G, Haendler S, Planes N, Josse E, Dimitriadis C A and Ghibaudo G 2015 New LFN and RTN analysis methodology in 28 and 14 nm FD-SOI MOSFETs *Proc. Annual International Reliability Physics Symp.—IRPS15* (New York: IEEE) pp XT.1.1–1.6

[115] Banaszeski da Silva M, Tuinhout H, Zegers-van Duijnhoven A, Wirth G I and Scholten A 2014 A physics-based RTN variability model for MOSFETs *Tech. Dig. Int. Electron Devices Meeting—IEDM14* (New York: IEEE) pp 848–51

[116] Sato T, Awano H and Hiromoto M 2014 A scalable device array for statistical device-aging characterization *Proc. ICSICT 2014* (New York: IEEE) pp 255–8

[117] Weckx P, Kaczer B, Chen C, Franco J, Bury E, Chanda K, Watt J, Roussel Ph J, Catthoor F and Groeseneken G 2014 Characterization of time-dependent variability using 32 k transistor arrays in an advanced HK/MG technology *Proc. Annual International Reliability Physics Symp.—IRPS15* (New York: IEEE) pp 3B1.1–1.6

[118] Kerber A 2014 Impact of RTN on stochastic BTI degradation in scaled metal gate/high-κ CMOS technologies *Proc. Annual International Reliability Physics Symp.—IRPS15* (New York: IEEE) pp 3B3.1–3.6

[119] Chen J, Tanamoto T, Noguchi H and Mitani Y 2015 Further investigation on traps stabilities in random telegraph signal noise and the application to a novel concept physical

unclonable function (PUF) with robust reliabilities *Symp. on VLSI Technol. Dig. of Techn. Papers* (New York: IEEE) pp T40–T41

[120] Hung K K, Ko P K, Hu C and Cheng Y C 1990 A unified model for the flicker noise in metal–oxide–semiconductor field-effect transistors *IEEE Trans. Electron Devices* **37** 654–65

[121] Hung K K, Ko P K, Hu C and Cheng Y C 1990 A physics-based MOSFET noise model for circuit simulation *IEEE Trans. Electron Devices* **37** 1323–33

[122] Martin S T, Li G P, Worley E and White J 1997 The gate bias and geometry dependence of random telegraph signal amplitudes *IEEE Electron Device Lett.* **18** 444–6

[123] Martin S, Li G P, Guan H, D'Souza S, Matloubian M, Claudius G and Compton C 1998 BSIM3 based RTS and 1/f noise models suitable for circuit simulations *Tech. Dig. Int. Electron Devices Meeting—IEDM98* (New York: IEEE) 95–88

[124] Martin S, Li G P, Guan H and D'Souza S 2000 A BSIM3-based flat-band voltage perturbation model for RTS and 1/*f* noise *IEEE Electron Device Lett.* **21** 30–3

[125] Leyris C, Pilorget S, Marin M, Minondo M and Jaouen H 2007 Random telegraph signal noise SPICE modeling for circuit simulation *Proc. ESSDERC-ESSCIRC 2007* (New York: IEEE) pp 187–90

[126] Brusmarello L, Wirth G I and da Silva R 2009 Statistical RTS model for digital circuits *Microelectron. Reliab.* **49** 1064–9

[127] Tang T B, Murray A F and Roy S 2010 Methodology of statistical RTS analysis with charge-carrier trapping models *IEEE Trans. Circ. and Syst.* **57** 1062–70

[128] Ashraf N, Vasileska D, Wirth G and Srinivasan P 2011 Accurate model for the threshold voltage fluctuation estimation in 45-nm channel length MOSFET devices in the presence of random traps and random dopants *IEEE Electron Device Lett.* **32** 1044–6

[129] Ashraf N and Vasileska D 2011 Static analysis of random telegraph noise in a 45-nm channel length conventional MOSFET device: threshold voltage and ON-current fluctuations *IEEE Trans. Nanotechnol.* **10** 1394–400

[130] Nour M A, Rouf A S M S, Çelik-Butler Z, Hou F C, Tang S, Sonnet A and Mathur G 2015 A scalable random telegraph signal simulation based on experimentally-identified gate oxide defects *Proc. of the 2015 Int. Conf. on Noise and Fluctuations—ICNF* (New York: IEEE) pp 1–4

[131] Nour M, Çelik-Butler Z, Sonnet A, Hou F-C, Tang S and Mathur G 2016 A stand-alone, physics-based, measurement-driven model and simulation tool for random telegraph signals originating from experimentally identified MOS gate-oxide defects *IEEE Trans. Electron Devices* **63** 1428–36

[132] Kaczer B *et al* 2011 Atomistic approach to variability of bias-temperature instability in circuit simulations *Proc. Annual International Reliability Physics Symp.—IRPS11* (New York: IEEE) pp 915–9

[133] Ayala N, Martin-Martinez J, Rodriguez R, Nafria M and Aymerich X 2012 Unified characterization of RTN and BTI for circuit performance and variability simulation *Proc. Europ. Solid State Device Research Conf.—ESSDERC '12* (New York: IEEE) pp 266–9

[134] Subirats A, Garros X, El Husseini J, Vincent E, Reimbold G and Ghibaudo G 2015 Modeling the dynamic variability induced by charged traps in a bilayer gate oxide *IEEE Trans. Electron Devices* **62** 485–92

[135] Higashi Y, Momo N, Sasaki Momose H, Ohguro T, Mitani Y, Ishihara T and Matsuzawa K 2014 Unified transient and frequency domain noise simulation for random telegraph noise and flicker noise using a physics-based model *IEEE Trans. Electron Devices* **61** 4197–203

[136] Bukhori M F, Roy S and Asenov A 2010 Simulation of statistical aspects of charge trapping and related degradation in bulk MOSFETs in the presence of random discrete dopants *IEEE Trans. Electron Devices* **57** 795–803

[137] Amoroso S M, Gerrer L, Markov S, Adamu-Lema F and Asenov A 2013 RTN and BTI in nanoscale MOSFETs: a comprehensive statistical simulation study *Solid-State Electron.* **84** 120–6

[138] Ye Y, Wang C-C and Cao Y 2010 Simulation of random telegraph noise with 2-stage equivalent circuit *ICCAD* (New York: IEEE) pp 709–13

[139] Luo H, Wang Y, Cao Y, Xie Y, Ma Y and Yang H 2012 Temporal performance degradation under RTN: evaluation and mitigation for nanoscale circuits *Proc. 2012 IEEE Computer Society Annual Symp. on VLSI* (New York: IEEE) pp 183–8

[140] Wirth G, Vasileska D, Asraf N, Brusamarello L, Della Giustina R and Srinivasan P 2012 Compact modeling and simulation of random telegraph noise under non-stationary conditions in the presence of random dopants *Microelectron. Reliab.* **52** 2955–61

[141] Camargo V V A, Kaczer B, Grasser T and Wirth G 2014 Circuit simulation of workload dependent RTN and BTI based on trap kinetics *Microelectron. Reliab.* **54** 2364–70

[142] Wang R *et al* 2013 A unified approach for trap-aware device/circuit co-design in nanoscale CMOS technology *Tech. Dig. Int. Electron Devices Meeting—IEDM13* (New York: IEEE) pp 834–7

[143] Luo M, Wang R, Wang J, Guo S, Zou J and Huang R 2014 Compact modeling of random telegraph noise in nanoscale MOSFETs and impacts on digital circuits *Symp. on VLSI Technol. Dig. of Techn. Papers* (New York: IEEE) pp 1–2

[144] Luo M, Wang R, Guo S, Wang J, Zou J and Huang R 2015 Impacts of random telegraph noise (RTN) on digital circuits *IEEE Trans. Electron Devices* **62** 1725–32

[145] Akyel K C, Ciampolini L, Thomas O, Turgis D and Ghibaudo G 2014 Impact of random telegraph signals on 6 T high-density SRAM in 28 nm UTBB FD-SOI *Proc. European Solid-State Device Research Conf.—ESSDERC14* (New York: IEEE) pp 94–7

[146] Aadithya K, Venogopalan S, Demir A and Roychowdhury J 2011 MUSTARD: a coupled, stochastic/deterministic, discrete/continuous technique for predicting the impact of random telegraph noise on SRAMs and DRAMs *Proc. 48th ACM/EDAC/IEEE Design Autom. Conf. (DAC)* (New York: IEEE) pp 292–7

[147] Aadithya K V, Demir A, Veugopalan S and Roychowdhury J 2013 Accurate prediction of random telegraph noise effects in SRAMs and DRAMs *IEEE Trans. Comput.-Aid. Design Integr. Circ. and Syst.* **32** 73–86

[148] Gerrer L, Ding J, Amoroso S M, Adamu-Lema F, Hussin R, Reid D, Millar C and Asenov A 2014 Modelling RTN and BTI in nanoscale MOSFETs from device to circuit: a review *Microelectron. Reliab.* **54** 682–97

[149] Pandey R, Saripally V, Kulkarni J P, Narayanan V and Datta S 2014 Impact of single trap random telegraph noise on heterojunction TFET SRAM stability *IEEE Electron Device Lett.* **35** 393–5

IOP Publishing

Random Telegraph Signals in Semiconductor Devices

Eddy Simoen and Cor Claeys

Chapter 6

RTS in memory and imager circuits

The first observations of a variable retention time in DRAM memories, corresponding to a bistability in the data retention time of a memory bit, were reported in the late 1980s and early 1990s [1, 2]. As pointed out in [3], it is believed that a possible origin for this is the generation of leakage current by a single defect in the silicon depletion region exhibiting metastable behavior. This is in fact quite similar to the RTS mechanism in CMOS imager circuits or in proton-irradiated CCDs. More recently, it has been demonstrated that threshold voltage fluctuations in flash memories are very sensitive to RTS noise, which is therefore a threat to their proper functionality [4]. In this chapter, we first discuss the issue of RTS in flash and SRAM memories. The latest results on RTS in DRAM and novel-concept memories (ReRAM or phase-change RAM) are then described. In a separate section, the occurrence of RTS in imager circuits will be briefly reviewed. For both types of circuits, dedicated analysis tools have to be developed in order to enable a statistical evaluation of the effect of RTS and its parameters.

6.1 RTS in flash and SRAM cells

The first reports on the effect of RTS in SRAM [5] and flash memories [6–8] appeared around the 90 nm CMOS technology node. It was, for example, concluded that RTS is higher in localized (channel hot-electron) programming than in uniform (Fowler–Nordheim) programming [6] and was created during the program/erase (P/E) cycling stress. This was explained by the fact that a non-uniform distribution of traps along the channel leads to an enhancement of the RTS. In addition, the RTS amplitude increases with programmed V_T state [6] and for a shorter gate length cell.

Among the digital devices, floating gate (FG) flash memories are considered to be the first that will be affected by RTS [4, 5]. This follows immediately from

doi:10.1088/978-0-7503-1272-1ch6

considering a simple number effect estimate of the influence of an RTS charge on the V_T spread [4]:

$$\Delta V_T = q/\left[W_{eff}L_{eff}C_{ox}R_c\right] \quad (6.1)$$

with R_c the coupling coefficient of the control to the FG. To achieve a low-cost bit in a flash memory, it is necessary to make the channel length and width almost equal to the feature size. C_{ox} cannot be enlarged because it is difficult to decrease the gate oxide thickness due to the SILC, which can be created during a read/write cycle. In addition, the stacked-gate structure of a Flash memory amplifies the V_T fluctuation by the coupling coefficient (<1). Moreover, the P/E cycles of a flash memory increase the number of interface or border traps, which introduce RTS fluctuations [4, 6]. The calculated effect on V_T is given in figure 6.1 [4], showing the 10 times worse effect of RTS on a flash memory compared to SRAM (a MOS transistor). Translating this data into a function of the process node demonstrates that the V_T shift due to RTS can reach more than 0.3 V in the 45 nm node (figure 6.2). In particular for multi-level Flash memory, this compromises the noise margins, as depicted in figure 6.3 [4].

Applying a statistical methodology on a large set of nMOSFETs demonstrates that the spread in V_T is larger at low drain currents than at higher I_D (figure 6.4) and that the distribution widens at lower operation temperatures (figure 6.5) [9]. These trends can be explained by the free carrier screening of the RTS becoming more efficient at higher currents or temperatures. At the same time, it was shown that the RTS time constants can be distributed over ten orders of magnitude (1 μs to 10^4 s), which points to trap variability in energy level and spatial location [9]. As outlined before, memory operation generally creates more traps with time, but sometimes trap removal (curing) can also be observed at room temperature [9].

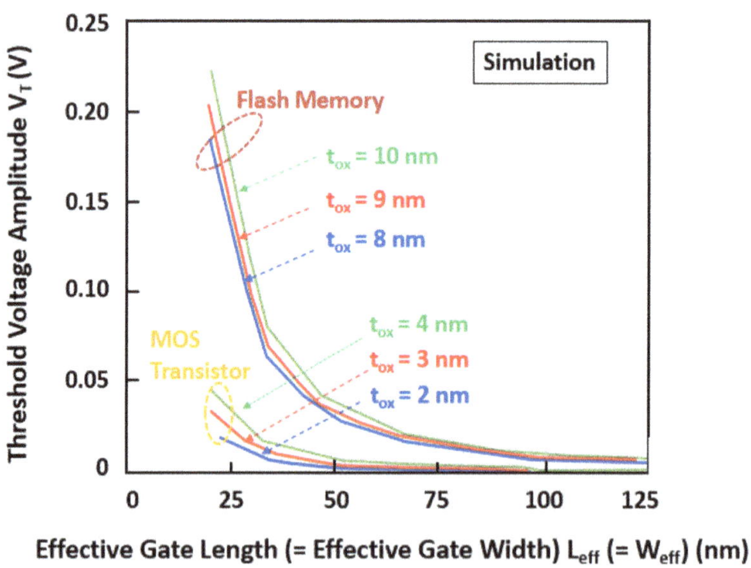

Figure 6.1. Calculation results of V_T amplitude as a function of L_{eff}. (After Kurata *et al* [4].)

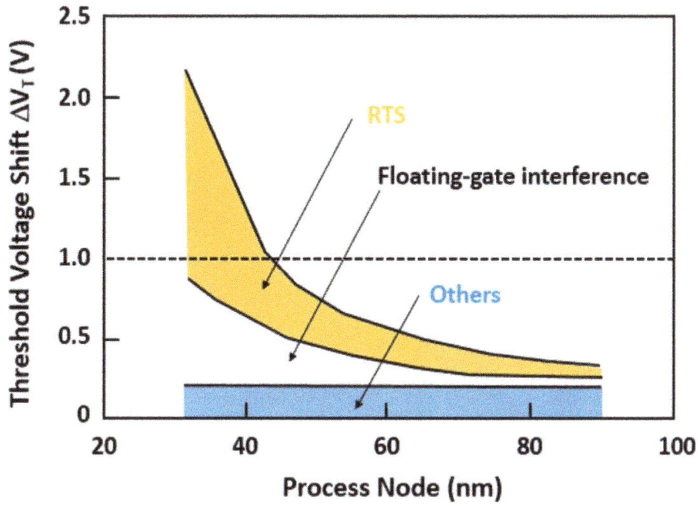

Figure 6.2. Estimation of V_T shift as a function of process node. (After Kurata *et al* [4].)

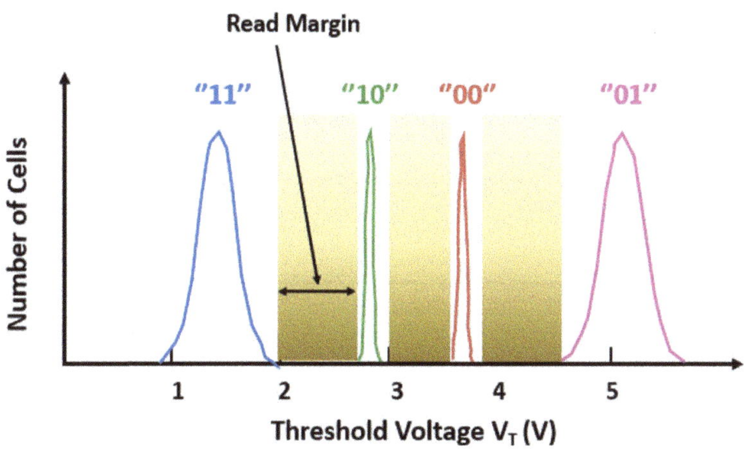

Figure 6.3. V_T distribution of multi-level flash memory. (After Kurata *et al* [4].)

The variability of the RTS energy level (or time constants) is illustrated in figure 6.6, showing a log-normal distribution for the mark space ratio [8]. The activation energy $E_a = E_T - E_F$ derived from an Arrhenius plot is shown to follow a normal distribution in figure 6.7 [8]. The width is about 1 eV and is enhanced by the presence of complex RTSs.

In recent years, numerous studies on statistical RTS-induced V_T fluctuations in flash memories have been reported (see e.g. [10–32]). The data have been analyzed using different types of distributions, such as log-normal [33]. The tail of large V_T fluctuations is certainly non-Gaussian and elongates with scaling. Interestingly, the RTS ΔV_T may become higher than the RDF around 22 nm and, therefore, constitutes a threat to SRAM stability (see figure 6.8) [33]. Other estimates put the limit at the 15 nm node [34].

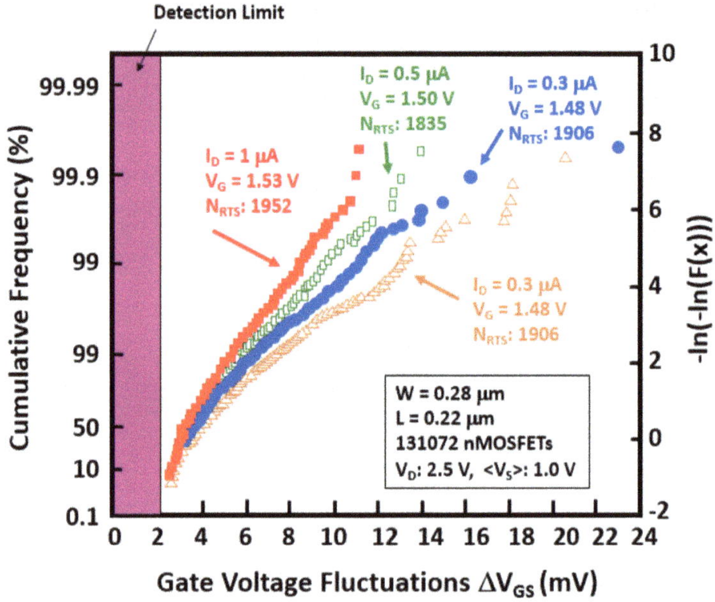

Figure 6.4. Drain current dependence of V_{GS} fluctuation (ΔV_{GS}). As the drain current decreases, the ratio of large ΔV_{GS} data increases. N_{RTS} indicates the number of extracted RTS samples from 131072 nMOSFETs in the array. (Reproduced with permission from Abe *et al* [9]. Copyright 2009 The Japan Society of Applied Physics.)

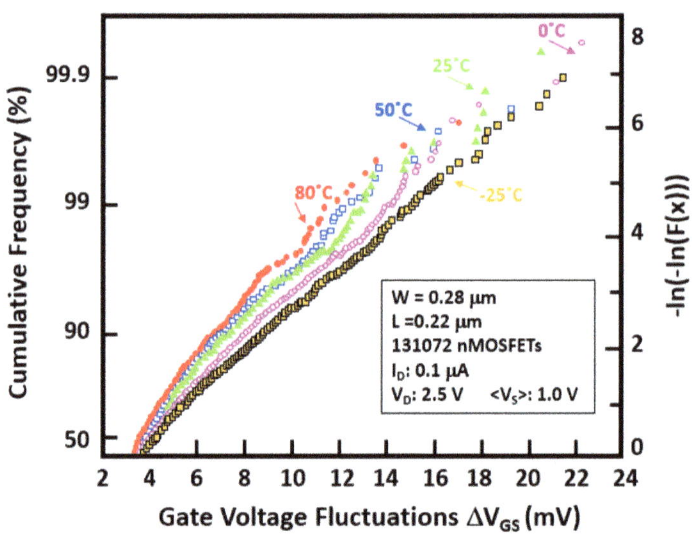

Figure 6.5. Temperature dependence of ΔV_{GS}. The tail of ΔV_{GS} decreases with increasing temperature. (Reproduced with permission from Abe *et al* [9]. Copyright 2009 The Japan Society of Applied Physics.)

A larger W_{eff} dependence than L_{eff} dependence was noted, pointing towards the importance of oxide defects at the isolation edges of the transistors; this is illustrated by the technology computer-aided design (TCAD) results in figure 6.9 [34]. This has been experimentally validated by comparing the noise factor in devices with and

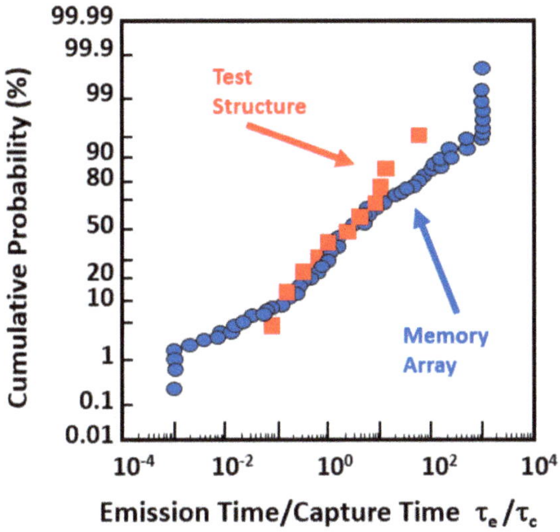

Figure 6.6. Log-normal distribution of τ_e/τ_c in the test structure (■) and memory array (●). (Reproduced with permission from Tega *et al* [8]. Copyright 2006 IEEE.)

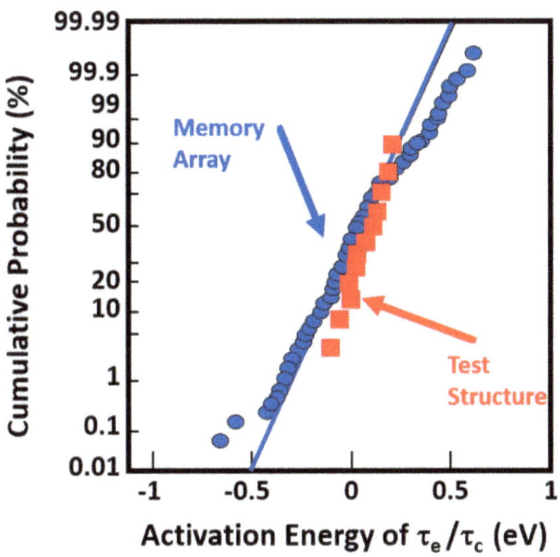

Figure 6.7. Normal distribution of activation energy in test structure (■) and memory array (●). (Reproduced with permission from Tega *et al* [8]. Copyright 2006 IEEE.)

without STI edges [35] and is in agreement with the interaction of traps with PPs. Detailed analysis has also revealed that nitrided oxides, while improving the cell reliability and endurance as well as data retention, can also increase the RTS noise, yielding a wider distribution [36].

In summary, the experimental V_T fluctuations associated with RTS are characterized in good approximation by a log-normal trend, represented by [37, 38]

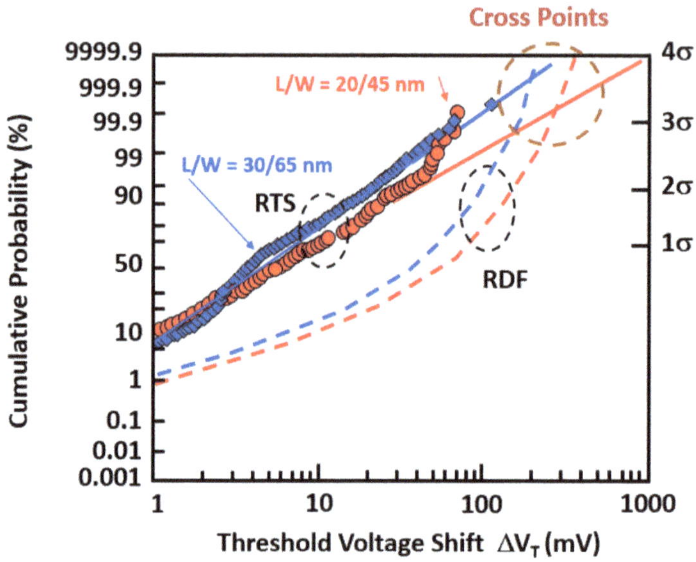

Figure 6.8. Distribution of RTS and RDF V_T variations in FETs with L/W = 20/45 and 30/65 nm, respectively. Projected RTS V_T variations exceed V_T variation at the 3 sigma level in 22 nm generation. (After Tega *et al* [33].)

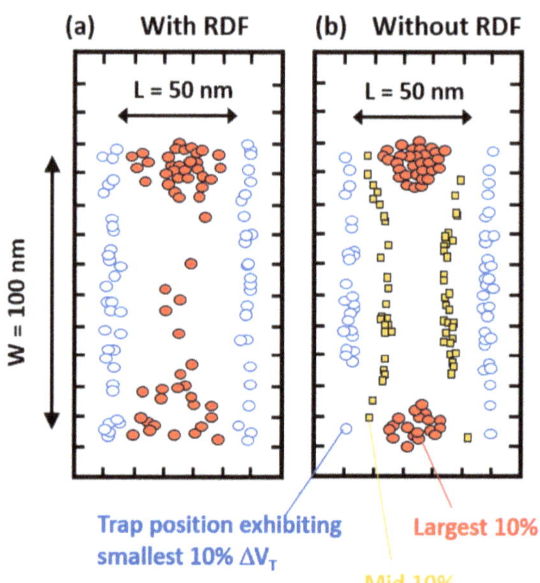

Figure 6.9. Trap position dependence of RTS amplitude (TCAD). The widthwise edge is sensitive to a charge due to potential dips formed by electric field crowding. (After Takeuchi *et al* [34].)

$$f(|\Delta V_\mathrm{T}|) = \frac{1}{\sigma}\exp\left(-\frac{|\Delta V_\mathrm{T}|}{\sigma}\right) \tag{6.2}$$

When plotted on a semi-log scale, the ΔV_T distribution is, therefore, a straight line whose slope λ_0 (mV dec^{-1}) is simply related to the distribution standard deviation σ by $\lambda_0 = \sigma\ \ln 10$ [37]. Using TCAD simulations based on an MC procedure, extensive investigation of the scaling trend of the statistical slope has been carried out [34, 37–41]. In good approximation, the scaling trend can be summarized as follows [37, 38]:

$$\lambda = \frac{K}{R_\mathrm{c}}\frac{t_\mathrm{ox}^{\delta}}{W}\frac{\sqrt{N_\mathrm{a}}}{\sqrt{L}} \tag{6.3}$$

where δ is a coefficient slightly lower than 1 and K is a constant. Moreover, calculations show that λ_0 decreases for higher V_DS, as there are fewer active traps, and at low V_DS traps near the S/D are less effective than mid-channel RTSs. Increasing V_DS shifts the distribution to the source, with fewer active traps, in agreement with the results of Asenov in figure 3.29 [42]. From equation (6.3), one derives that the scaling of the device width has a stronger impact than the length. Finally, it has been demonstrated that the experimental trend for NOR and NAND flash memories is correctly reproduced by the modeling, as shown in figure 6.10 [37, 38].

The effect of non-uniform channel doping has been investigated experimentally and using 3D atomistic simulations, showing that the distribution of the RTS amplitudes may deviate from a purely exponential one [43]. In order to reduce the RDF amplification effect, one can rely on retrograde and δ-shape doping. The benefits offered by such approaches increase by placing the high doping regions

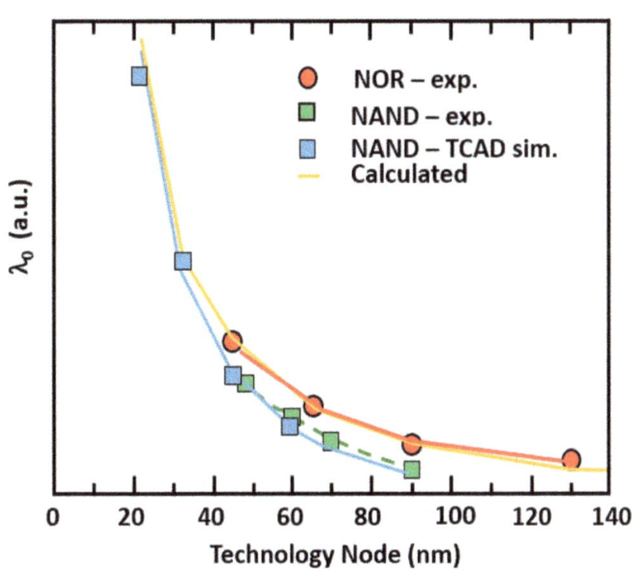

Figure 6.10. Measured and calculated λ_0 for NAND and NOR flash technologies. (Reproduced with permission from Ghetti *et al* [37]. Copyright 2008 IEEE.)

further away from the channel. It has been shown that the suppression of RTS by engineered doping profiles is correlated with a reduction in cell threshold voltage variability [43]. In addition, it has been shown that the cell shape plays a strong role in the RTS of deca-nanometer flash memories [44].

For the development of 3D stacked flash memory, tube-type poly-silicon channel transistors are being developed [45, 46]. Again, they exhibit RTS, which may interfere with memory operation [45–51]. The induced V_T shift is larger than can be expected from a simple charge sheet model and points to the percolative character of the conduction in a polycrystalline channel [47]. The dependence of the RTS amplitude on the type of material, i.e. the deposition conditions and the average grain size, indicates that the main origin of the traps is in the poly-Si channel. The effect of a single trap is shown to increase with average grain size [48]. Two main types of RTS have been identified [48–51], based on an analysis of the capture and emission time constant behavior versus gate voltage. A first type occurs preferentially at low currents (subthreshold regime), showing a strong dependence of the time constants on V_{GS} (large coupling factors), while a second type exhibits barely any gate voltage dependence, i.e. α_e and $\alpha_c \sim 0$ (figure 6.11) [49, 50]. This points to the fact that the first type of trap occurs mainly at the interface (or in the gate oxide), while the second type is present in the poly-Si channel. This is further supported by the fact that their occurrence is largely suppressed in a so-called 'macaroni'-type channel [50], where the central part of the channel is replaced by an oxide so that only a thin conductive shell remains. This removes most of the traps in the material and, hence, the RTS [50].

A similar statistical analysis has been performed for scaled-down SRAM cells [52, 53]. Overall, it was found that the RTS-induced V_T fluctuations are more pronounced for p-channel compared to n-channel devices. This has been ascribed to the larger hole capture cross section, yielding one decade shorter capture times, and to the mobility fluctuations contribution to the RTS amplitude, which appears to be the dominant enhancement factor [52]. Using an alternating-bias RTS characterization technique, which shortens the measurement time by a factor of 10, has revealed that the corresponding write margin degradation is less than 50 mV in a 45 nm SRAM technology [54]. There exists, however, concern about the stability of SRAM cells suffering from RTS, as pointed out by some recent studies [55–58].

6.2 RTS in DRAM and logic circuits

In the past, a clear link has been established between the variable retention time of DRAM cells and RTS in the junction leakage current [1, 2]. One of the sources of junction leakage is defects in the depletion region (silicon substrate), which could thus also be responsible for the RTS when strategically placed. This has in fact been demonstrated for the drain-to-substrate junction in advanced CMOS devices [59], indicating that the RTS amplitude increases with the substrate doping density, i.e. the electric field in the depletion region. This means that TAT may play a role in the carrier transitions to and from the defect. A detailed study on transistors for DRAMs, combining different analytical methods, has successfully identified one

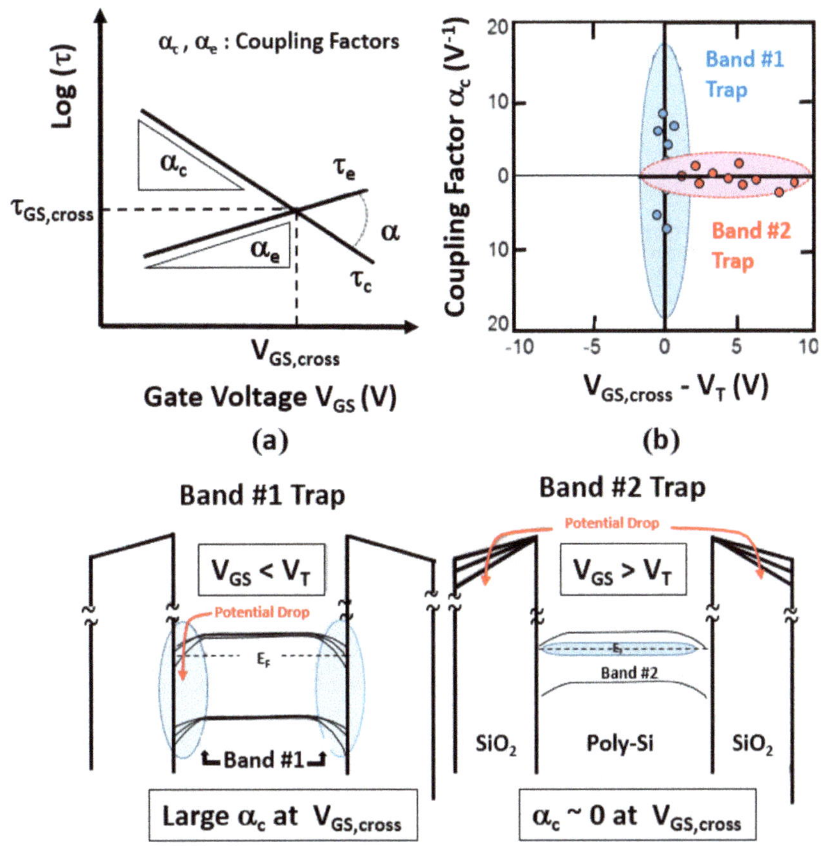

Figure 6.11. (*a*) Definition of $V_{GS,cross}$, α_c and α_e. (*b*) For band #1 traps, the capture probability increases with V_{GS} (large α_c). For band #2 traps, the capture probability is roughly constant with gate voltage (α_c is insensitive to V_{GS} and ~0). (*c*) For band #1 traps, E_F sweeps the poly-Si bandgap progressively charging defects at the interface and in the poly-Si bulk. (*d*) For band #2 traps, E_F is pinned in the channel and only traps aligned with E_F exhibit RTN. They can be in the bulk or near the interface. (Reproduced with permission from Lee *et al* [50]. Copyright 2015 Elsevier.)

possible type of defect, which can be responsible for variable retention times [60, 61]. It has been shown that a metastable V_2O_x complex, schematically represented in figure 6.12, could be the source of RTS. The metastability comes from the energy-level splitting under the influence of strain, whereby one defect orientation becomes more metastable (0.1 eV higher in energy) than the defect aligned with the main strain direction. An activation barrier $E_{g,b}$ of 1.4 eV has to be surpassed to flip the defect from one orientation to the other. In the 'bad' high-leakage state, the center becomes more effective in leakage current generation, thereby reducing the retention time. As in irradiated CCDs, it is believed that the carrier generation rate is enhanced by high electric fields [60]. It has been shown that defect engineering, preventing the formation of the complexes, is the most effective in removing the variable retention time bits in DRAM circuits [61].

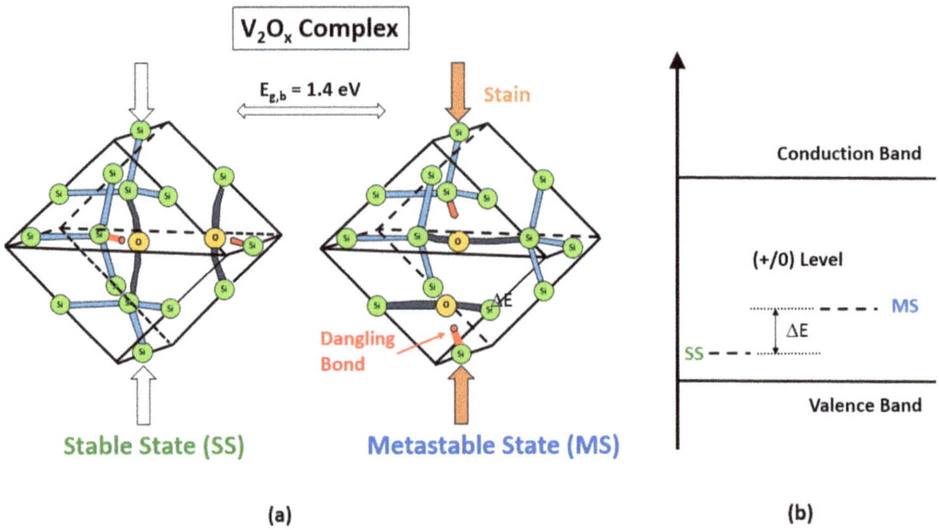

Figure 6.12. Model for a bistable V_2O_x defect under strain: (*a*) atomic view and (*b*) energy-level view. (After Umeda *et al* [60].)

However, recent evidence points to the fact that oxide traps giving rise to RTS in the GIDL current may also be partly responsible for the variable retention in advanced DRAM cells [59, 62–66].

While it is clear that NVM and other memory types are the first to be affected by V_T instabilities induced by RTS, there is also concern regarding logic circuits [67]. This topic has been studied in 65 nm CMOS ring oscillators, showing that the effect is only 4% within die variation (WID) on delay. However, as seen in figure 6.13, the effect becomes more pronounced for scaled technologies and scales faster, i.e. according to $1/WL$ than the RDF, which follows Pelgrom's law ($\sim 1/\sqrt{WL}$). This means that one should implement RTS aware design for logic circuits as well [67].

6.3 RTS in novel ReRAM and PCMs

There is currently strong interest in the development of so-called ReRAM (or RRAM) for the next generation of NVM applications. The advantages of ReRAM are its superior scalability, fairly simple processing, compatibility with standard CMOS, low-voltage operation, fast P/E speed and superior endurance. The memory operation is based on the gate-bias-induced change of the resistance of a metal–insulator–metal structure between a high (R_H) and a low (R_L) value, as shown in figure 6.14 [68, 69]. Moreover, if bipolar switching is possible, as in figure 6.14, this will lead to a high read disturb immunity. In addition to Ta_2O_5/TiO_2 [68, 69], TiN/$TiON/SiO_2$ [70–72], $Al/Pr_{0.7}Ca_{0.3}MnO_3$ [73], NiO [74, 75], $Al/TiO_x/Al$ [76–79], ZrO_x/HfO_x [80], AlO_x/WO_y [81] and HfO_x-based stacks [82–91] are among the dielectric stacks under investigation for this kind of application. However, as for other memory types, the occurrence of RTS can cause retention loss and threatens the stability of the memory and should, therefore, be studied in detail.

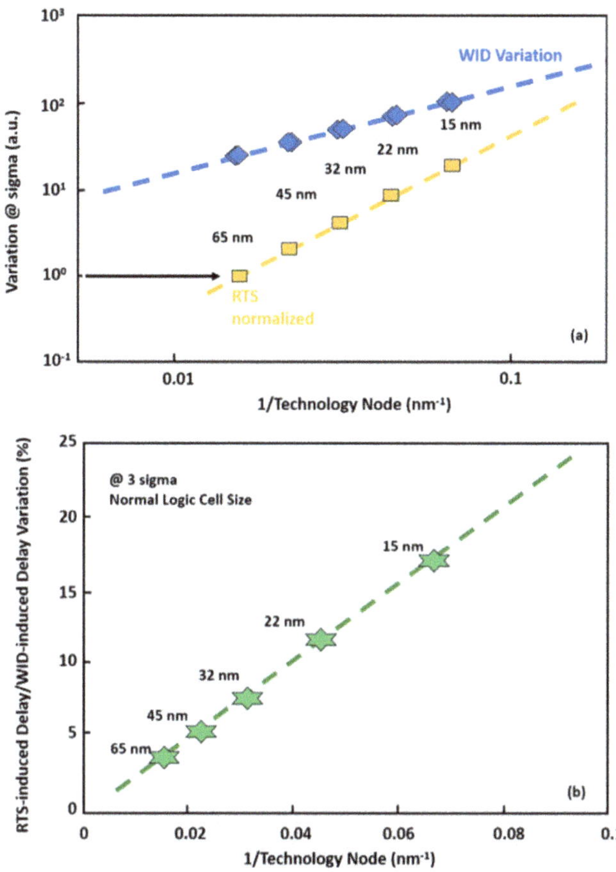

Figure 6.13. (*a*) Predicted RTS-induced delay fluctuation and WID variation-induced delay variation in a CMOS logic circuit, and (*b*) ratio of the two variations indicating that in more scaled processes, the effect of RTS will become visible on the circuit delay. (After Ito *et al* [67].)

From the first reports in the literature, it is clear that the presence of defects (mainly oxygen vacancies) in the proximity of the current conduction path of the resistor will modulate the charge transport and give rise to RTS. The formation of the defects strongly depends on the details of the gate stack (processing). It is also evident that a higher $1/f$ or RTS noise is generally found in the high-resistance state (HRS) compared to the low-resistance state (LRS), usually exhibiting predominantly $1/f$-like noise (figure 6.15) [77].

Several qualitative models have been advanced to explain the occurrence of RTS in the HRS of ReRAM structures, which largely depend on the dominant transport mechanism. In the case of bipolar TiN/α-TiO_x/TiN stacks, it was shown that the RTS-related traps are near the top electrode, where Ti scavenges (negative) oxygen from the dielectric layer, forming oxygen vacancies [77]. The alignment of positively charged oxygen vacancies forms a conductive path (figure 6.16) [80], which can be reoxidized during a reset bias pulse. This is a pulse at sufficiently high bias and

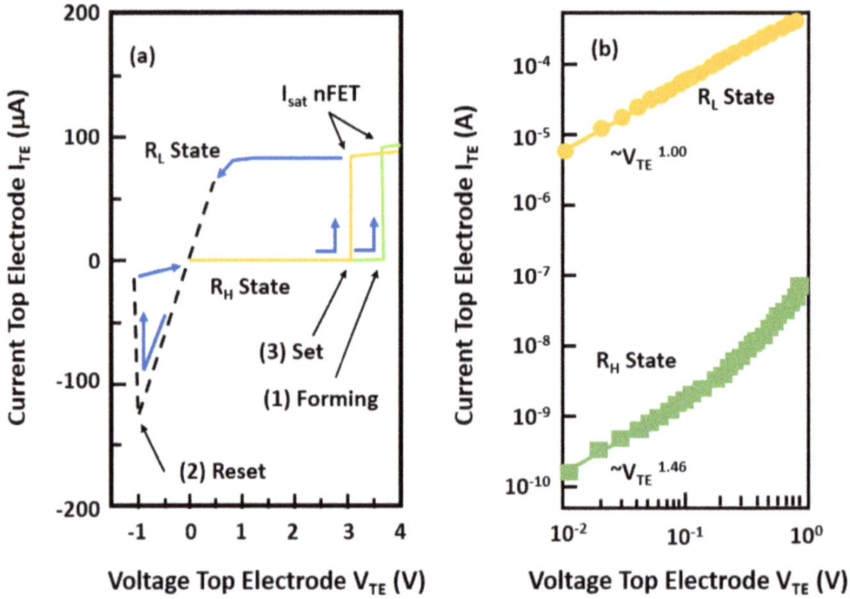

Figure 6.14. (*a*) Typical DC switching characteristics and (*b*) *I–V* curves in the R_L and R_H states for a one-transistor/one-resistor type Ru/Ta_2O_5/TiO_2/Ru ReRAM, (After Terai *et al* [69].)

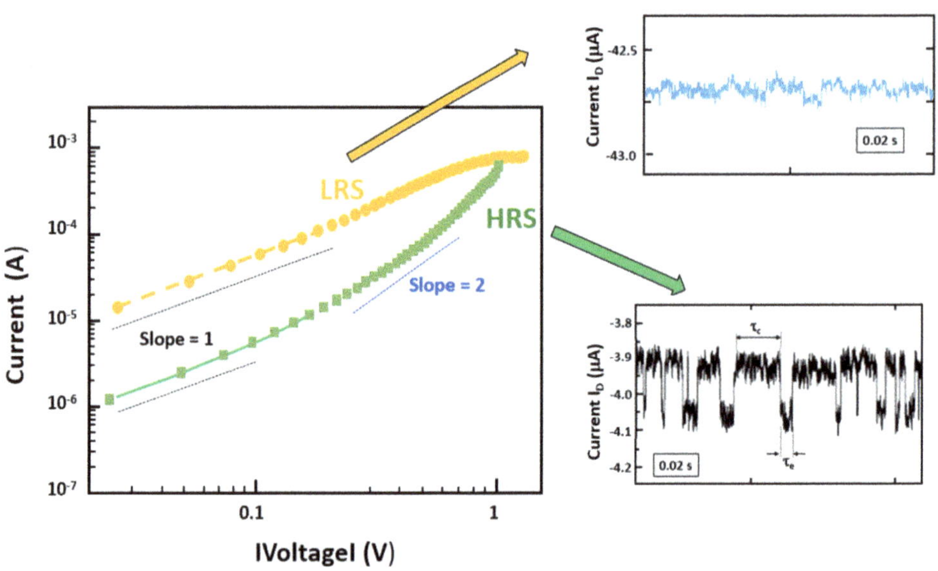

Figure 6.15. Logarithmic plot of the *I*–|*V*| curve for the negative voltage region. The upper and lower insets represent current fluctuations of the sample with an area of 100 × 66 nm^2 in the LRS and HRS, respectively, at the voltage of −0.08 V. (After Lee *et al* [77].)

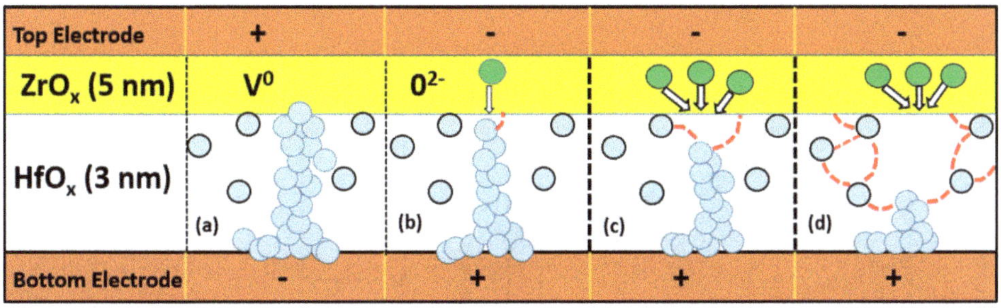

Figure 6.16. Proposed model based on noise analysis. (*a*) The conducting path is generated by applying the positive SET voltage during LRS. (*b*) The conducting path is oxidized by oxygen migration due to the application of the RESET voltage. (*c*) and (*d*) The oxidized region expands as the negative RESET voltage is increased. (After Lee *et al* [80].)

current to produce sufficient heat for migration of oxygen ions (or oxygen vacancies) to occur. It is generally assumed that in the HRS, the current (or percolation) path is broken (or constricted), which is at the origin of the high resistance and the RTS [71, 76, 78, 86, 87].

According to figure 6.17 [80], the statistical spread in RTS amplitude ΔI decreases for smaller area structures, as the total number of traps decreases in the same way as the PSD. In this case (ZrO_x/HfO_x stack), the HRS becomes more stable for smaller area, owing to the trap-induced instability of the parameters. It was also found that for a higher reset voltage, a higher RTS and $1/f$ noise were obtained, indicating more traps being formed [79].

While RTS in RRAM is clearly trap related, different scenarios have been proposed for its origin. It is assumed either that a trap adjacent to the conductive path communicates and modulates the current through emission/capture (RTS in LRS) [79] or that the current is carried by a single trap close to the gate electrode in a constricted filament, which could cause the high RTS in the HRS [86, 87]. Recently, a new model for the switching variability of HfO_x-based RRAM has been proposed [92], considering the statistical distribution of ion migration parameters. The model accounts for size-dependent RTS, while bias-dependent RTS is explained by Poole–Frenkel transport and Joule heating in atomic-sized current filaments. Based on this, a temperature sensor using a contact RRAM has been proposed, relying on the relationship between an RTS and temperature [93].

More recently, a detailed numerical model has been derived for RTS-induced by a trap adjacent to the channel, but not participating in the conduction, i.e. not part of the conducting filament (CF) [75, 94, 95]. The basic idea is sketched in figure 6.18. According to figure 6.19, the normalized RTS amplitude ($\Delta R/R$) increases with resistance up to 100 kΩ and then saturates [95]. This can be explained based on size-dependent carrier depletion in the CF by the charged trap. For a narrow filament with a small diameter ϕ, the Debye length λ_D has a value high enough to completely deplete the filament and high RTS results.

Based on this concept, a numerical model was developed including Joule heating [95]. The latter gives rise to a shortening of the defect switching time, since the time

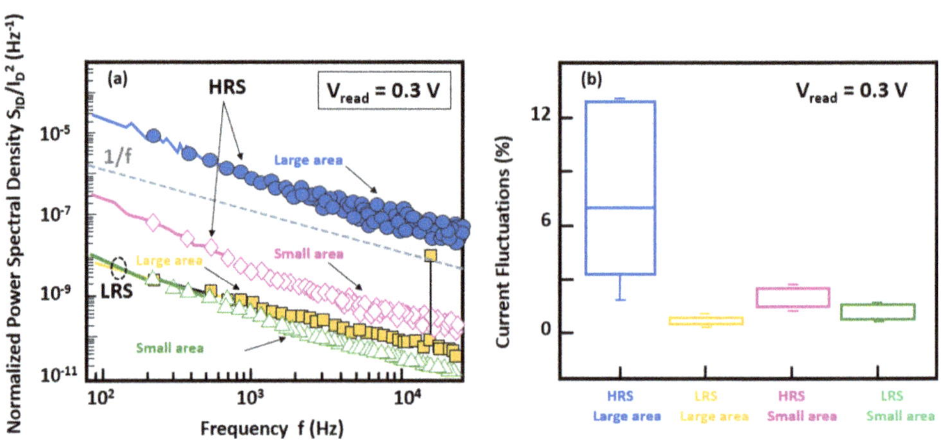

Figure 6.17. (*a*) The PSD value of small-area devices is more than two orders of magnitude lower than for large-area devices for HRS, which indicates the reduced influence of defects. (*b*) Statistical ΔI decreases with the scaling down of the device area. This result indicates that the current flow becomes more stable with the reduced influence of defects with the scaling down of the device area. The average values of the current for HRS and LRS in large scale devices are 0.1 and 0.8 mA, respectively, at a gate bias of 0.3 V. In the small-area devices, the average values are 0.05 and 0.6 mA, respectively. (After Lee *et al* [80].)

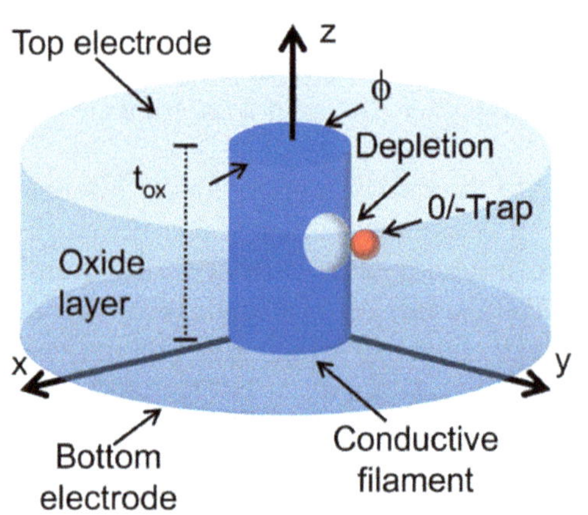

Figure 6.18. Schematic picture of the RTN mechanism due to a defect with fluctuating charge at the CF surface. The CF diameter ϕ and the CF height t_{ox}, equal to the oxide thickness, are shown. When the defect is negatively charged, the CF is depleted from carriers within approximately a Debye length. The reduced defect concentration causes a transition to the higher resistance level. (Reproduced with permission from Ambrogio *et al* [95]. Copyright 2014 IEEE.)

constants are thermally activated [94]. The doping density N_D in figure 6.19 reflects the local concentration of defects (oxygen vacancies most likely—V_O), acting as n-type dopants in transition metal oxides. It is considered that the total image charge induced by the trap is $+q$, which is comparable to the number of charges present in the CF [95]. The proposed model can also describe complex, multi-level RTS, due to two independent traps. As a result, it has been shown that the amplitude of one trap

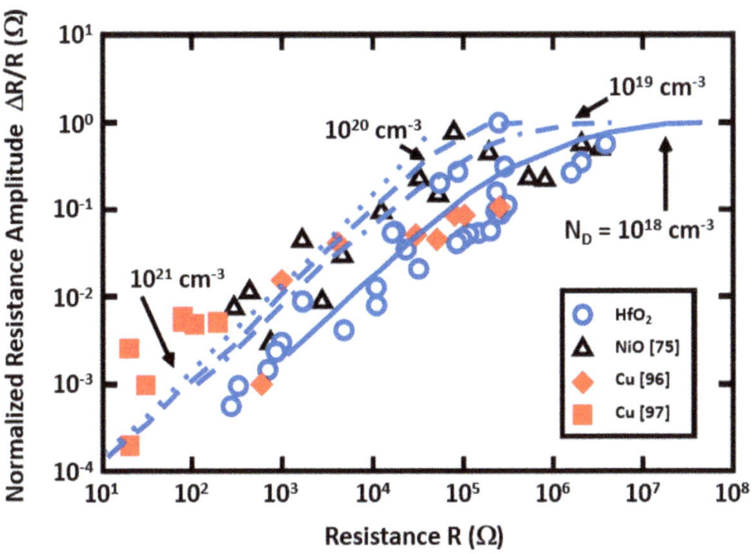

Figure 6.19. Measured and calculated $\Delta R/R$ as a function of R. Data for NiO-based RRAM [75], Cu-based electrochemical RRAM [96] and Cu nanobridges [97] are also shown for comparison with HfO_x-RRAM data. All data display a universal dependence on R, which can be understood by size-dependent depletion. Calculations were carried out at variable doping density N_D. (Reproduced with permission from Ambrogio *et al* [95]. Copyright 2014 IEEE.)

depends on the state of the other. This is due to the partial overlap of the depletion region of the two traps in the negative charge states [95]. Moreover, it has been shown that the off and on times and the amplitude decrease with increasing read voltage (V_{read}) or current [96]. This is related to the Joule heating induced by the dissipated power in the CF. This enhances the local temperature at the defect site. Based on the fact that these changes only depend on the absolute value of V_{read} supports Joule heating as the root cause and not defect migration. If this were the case, the variation would be irreversible and in one direction.

It was concluded from the symmetric and non-linear behavior of resistance with bias that Poole–Frenkel conduction, depicted in figure 6.20, dominates the transport in the CF. Assuming Poole–Frenkel conduction in the filament, a simple expression for the RTS amplitude has been derived [95], namely

$$\frac{\Delta R}{R} = \left[\exp\left(\frac{\Delta E_{hop}}{kT}\right) - 1\right] \tag{6.4}$$

with E_{hop} the characteristic energy barrier for electron hopping and ΔE_{hop} the increase of this barrier by the charging of the trap (figure 6.21). Based upon the model, guidelines for improving RRAM operation have been formulated, whereby reading at relatively large V_{read} should make the device less sensitive to RTS. The optimum V_{read} is then a trade-off between RTS reduction and read disturb [95].

A different modeling approach has been followed by the Modena Group (Italy), who performed a detailed analysis of the RTN of HfO_2-based RRAM cells in the

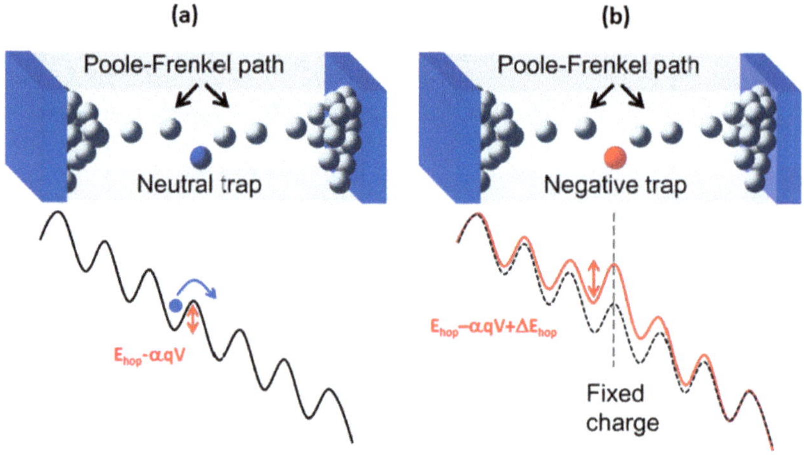

Figure 6.20. Schematic illustration of the Poole–Frenkel transport in a few-defect CF. For a neutral defect, R is controlled by an energy barrier E_C (*b*) while the transition to a negative charge causes an increase of the energy barrier by ΔE_C. (*a*) Neutral trap. (*b*) Negative trap. (Reproduced with permission from Ambrogio *et al* [95]. Copyright 2014 IEEE.)

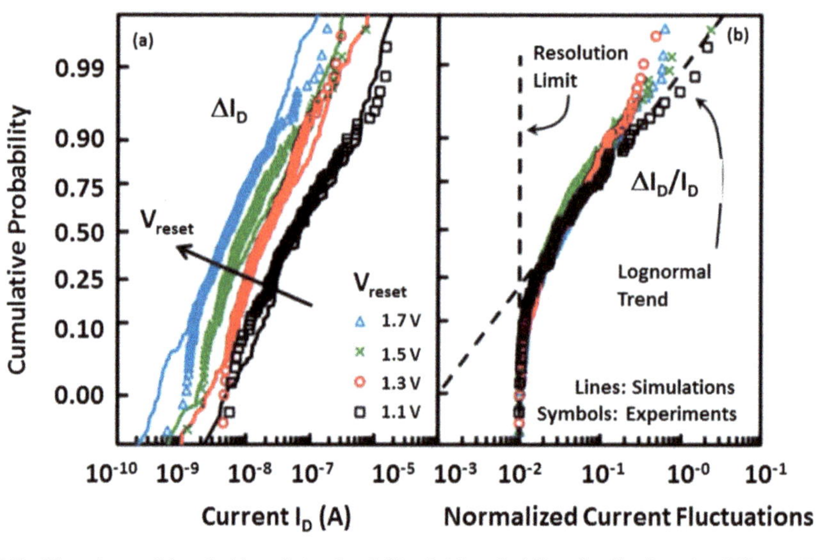

Figure 6.21. Experimental (symbols) and simulated (lines) ΔI probability distributions for different V_{reset}. The constant slope of the ΔI distribution indicates that the defect distribution in the barrier is uniform and its density is unaffected by V_{reset}. (Reproduced with permission from Puglisi *et al* [101]. Copyright 2015 IEEE.)

HRS [83, 84, 98–101] and in the LRS [102, 103]. In order to enhance the sensitivity of the measurements, a factorial hidden Markov model (FHMM) analysis of the time current fluctuations has been developed [101], enabling a more accurate separation of the different RTS components in complex signals, which typically occur in the HRS. An important parameter, defining the cross section of the CF, is the compliance limit during the forming (and the set operation) [84]. The filamentary conduction in the LRS usually behaves in an ohmic manner and is carried by a high density ($\sim 10^{21}$ cm^{-3}) of defects—typically V_O^+ centers. The HRS is formed by a reoxidation of the bottom part of the CF, reducing the density of oxygen vacancies locally and creating an insulating barrier for conduction. In this thin barrier layer, conduction is believed to occur through multi-phonon TAT (MPTAT) [98–101]. The thickness of the barrier layer is defined by the magnitude of the reset compliance. Assuming that RTS is produced by positively charged oxygen vacancies in the barrier, assisting the TAT conduction of the HRS filament, simple expressions for the capture and emission time constant can be established [84]:

$$\tau_c^{-1} = c_0 N_C \exp\left(-\frac{x_T}{\lambda_C}\right)\exp\left(-\frac{E_{cap}}{kT}\right) \tag{6.5a}$$

$$\tau_e^{-1} = c_0 N_C \exp\left(-\frac{t_{ox} - x_T}{\lambda_E}\right)\exp\left(-\frac{E_{em}}{kT}\right) \tag{6.5b}$$

with x_T the trap distance from the (negative) cathode), λ_C and λ_E the characteristic tunneling depths calculated through the WKB method neglecting the electric field in the barrier, and N_C the density of states at the bottom of the CB. The Fermi–Dirac occupation probability is assumed to be equal to 1 at the cathode and zero at the anode. E_{cap} and E_{em} are the associated thermal activation barriers and they depend on the trap position, relaxation energy and applied bias [85]. They are given by [101]

$$E_{cap} = \frac{\left[E_{relax} - \left(E_T - \varphi_B + qV_{read}\frac{x_T}{t_b}\right)\right]^2}{4E_{relax}} \tag{6.6a}$$

$$E_{em} = \frac{E_{relax}}{4} \tag{6.6b}$$

where E_T is the trap ionization energy, E_{relax} is the relaxation energy, t_b is the barrier thickness and ϕ_B is the energy offset at the electrode/oxide interface. The second exponential in the equations takes into account the lattice rearrangement required for accommodating (removing) the electron charge into (from) the trap during a capture (emission) event. The effective position within the dielectric barrier in the HRS can then be derived from [84]

$$z_{eff} = \frac{\partial\left(\ln\left(\frac{\tau_c}{\tau_e}\right)\right)}{\partial V_{read}}\left[\frac{1}{2}(\delta + \varphi V_{read})\right] \approx \frac{\partial\left(\ln\left(\frac{\tau_c}{\tau_e}\right)\right)}{\partial V_{read}}\frac{1}{2}\delta \tag{6.7}$$

The term in parentheses accounts for the structural relaxation occurring during the trapping/detrapping processes. Typical values for the relaxation energy of V_O^+ in HfO_2 are in the range of 1.19 eV [84]. Using equation (6.6), it has been found that the dominant RTSs are at 2/5 of the dielectric barrier. For a higher reset value, more RTSs (and a thicker barrier) are revealed, corresponding with a smaller current.

Ultimately, for a high reset voltage, a large number of RTSs are created resulting in a $1/f$ noise spectrum [84].

Further analysis of RTS in the HRS has revealed that there is no device area dependence of the statistical amplitude distribution, indicating the localized filamentary nature of the charge transport [99–101]. A log-normal distribution is typically observed for the RTS amplitude ΔI (figure 6.21(*a*)), due to the normal variations of the dielectric barrier thickness. The ΔI distribution shifts rigidly with V_{reset} since the average I_{HRS} decreases due to the increase of the barrier thickness [101]. It reduces exponentially with V_{reset} due to the reduction of the tunneling probability. In contrast, $\Delta I/I$ is not affected by V_{reset} in figure 6.21(*b*), showing the strong statistical correlation between ΔI and I [101]. This *nearly excludes the depletion by an adjacent trap model* of figure 6.18 to explain these observations [101]. In addition, the RTS distribution shifts to higher values with temperature, consistent with the strong T dependence of the TAT current. Again, $\Delta I/I$ is not affected by the temperature [101]. Also the capture and emission times follow log-normal distributions (figure 6.22) [101].

Comparing RTS to the cycling variability in the HRS (reset–read–set–read–reset) no direct correlation has been found [101]. The cycling variability results from the intrinsic randomness of the reset process, which affects the properties of the CF in the HRS (thickness and defect distribution in the oxidized tip near the cathode). The number of defects found by the FHMM algorithm shows some random variations at each cycle. Two distinct physical mechanisms are responsible for I_{HRS} variability and RTS: δI_{HRS} is mainly due to variations of the barrier thickness and redistribution of TAT supporting defects over cycling, while RTN is associated with the

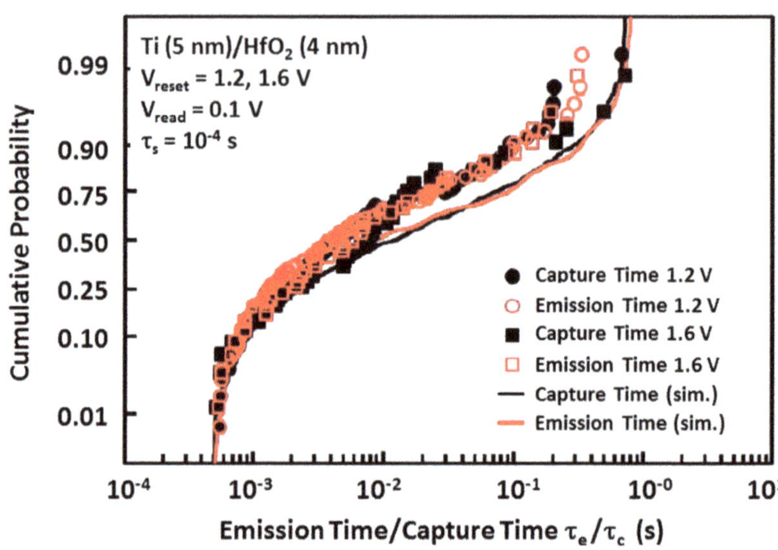

Figure 6.22. Experimental (symbols) and simulated (lines) distributions of the capture and emission time. Capture and emission time probability distributions for different V_{reset} show a good agreement (overlap). (Reproduced with permission from Puglisi *et al* [101]. Copyright 2015 IEEE.)

random activation/de-activation of the TAT supporting defects within the barrier [99].

According to figure 6.23, the capture time increases exponentially with V_{read}, which indicates that a higher energy is released by electrons to the lattice at every capture event. On the other hand, τ_e is approximately constant; it corresponds with electron emission from the ground state of the trap not affected by V_{read}. At the same time, τ_c may either increase or decrease depending on the energy misalignment between electrons and the defect. From the extracted experimental E_{relax} (1.8–2.7 eV) it is clear that V_O^+ yields too low a value and cannot be responsible for RTS itself, while it does assist the conduction across the dielectric barrier, albeit with much smaller capture and emission time constants (see below). This implies that a different slow trap species may be responsible for the RTS, with as a possible candidate neutral interstitial oxygen, for example [101]. This follows from DFT modeling of defect states in HfO_2. Alternatively, the switching of the metastable current-carrying V_O^+ state to a more stable charge state (V_O^{2+}?) where it is not active can also explain the observed RTS fluctuations [101]. These two possible mechanisms are depicted in figure 6.24 [101] and have been proposed by other research groups as well [104].

Finally, according to figure 6.25, there exists a kind of universal relationship between the effective number of (RTS) traps ENOT—to be considered a figure-of-merit of the statistical complexity of the current fluctuations—and the barrier thickness extracted from the current in the HRS from the TAT model [101]. The linear relationship again emphasizes a more or less constant trap density in the dielectric barrier, independent of the stack composition and V_{reset}, while at the same time the complexity of the fluctuations increases with V_{reset} and t_b. Most probably, it is a property of the material at a given substoichiometric level. This suggests that materials with a low fractional oxygen content and/or a high lattice flexibility (the capability to arrange the lattice upon some defects creation) may constitute a preferential choice to limit the concentration of defects involved in RTS [101].

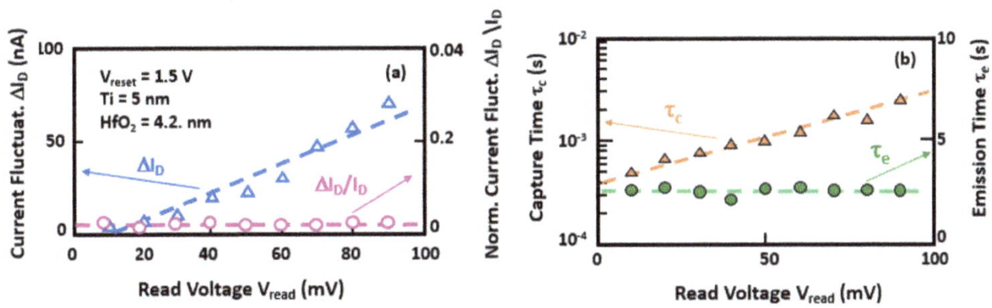

Figure 6.23. (*a*) Experimental ΔI (left *y*-axis) and $\Delta I/I$ (right *y*-axis) average values for different V_{read}. ΔI scales with the average current I. (*b*) Experimental capture (left *y*-axis) and emission (right *y*-axis) times average values for different V_{read}. The capture time exponentially depends on the read voltage, while the emission time shows no dependence, in agreement with the TAT model. (Reproduced with permission from Puglisi *et al* [101]. Copyright 2015 IEEE.)

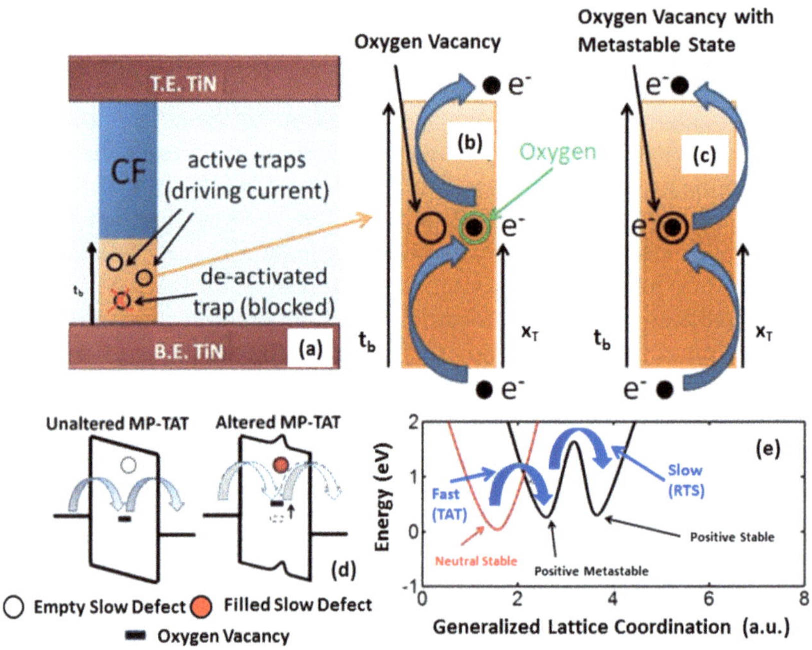

Figure 6.24. (*a*) Schematic of the device in HRS. Representation of two mechanisms possibly responsible for the RTS. The barrier thickness and the trap location are highlighted. (*b*) Charge trapping at a slow defect (in green, probably the O interstitial) may induce a Coulomb blockade effect on a close oxygen vacancy, normally assisting the TAT transport. (*c*) O vacancy activation and de-activation through a metastable state configuration. (*d*) Schematic of the Coulomb blockade. The charge trapping at the slow defect in the proximity of the O vacancy changes the potential and the energy alignment of the O vacancy, altering the TAT current. (*e*) Generalized coordination diagram illustrating the O vacancy model with metastable states. The TAT current involves the charge transition between the neutral and the positive metastable state. The RTS is generated by the relaxation of the positive metastable state into a stable configuration. This process temporarily inhibits the TAT transport and causes RTN. (Reproduced with permission from Puglisi *et al* [101]. Copyright 2015 IEEE.)

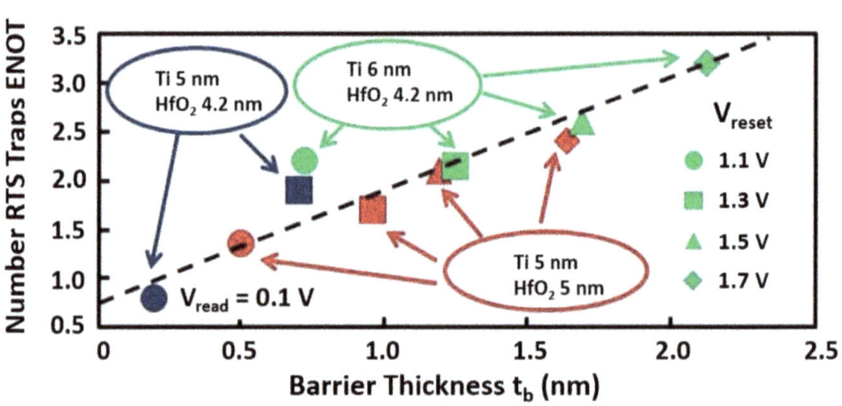

Figure 6.25. ENOT (symbols) versus average barrier thickness for different device stacks reset at different V_{reset} along with a linear fitting. Increasing barrier thickness results in increasing statistical complexity of RTN. (Reproduced with permission from Puglisi *et al* [101]. Copyright 2015 IEEE.)

Detailed studies of RTN and cycling variability in the LRS [102, 103] provide evidence for the depletion model represented in figure 6.18 [95]. Overall, the amplitude is at most 1% and, hence, much smaller than for the HRS and follows a log-normal distribution. This rather unexpected trend is related to the random distribution of the traps around the CF [102].

A microscopic physical description of RTN in HfO_x-based RRAM has been proposed, based on the two models of figure 6.24 [105]. As a result, it has been evidenced that the time constants involved in HRS TAT conduction assisted by V_O^+ defects in the dielectric barrier are 10 orders of magnitude higher than the RTS time constants (figure 6.26). Studies of the impact of LFN on read distributions of ReRAM cells *and arrays* have been reported [106–108]. A simple analytical model has been derived for the standard deviation in the read current I_R when RTS fluctuations are present in the ReRAM [107]. It is given by

$$\frac{\sigma_{I_R}}{I_R} = \frac{\sqrt{\gamma_1 - \gamma_2 \operatorname{atan}\left(\frac{2\pi\tau_0}{t}\right)}}{I_R} \tag{6.8}$$

with

$$\gamma_1 = \frac{2\Delta I_R^2 \tau_0}{\pi(\tau_c + \tau_e)} \operatorname{atan}\left(2\pi\tau_0 f_{max}\right) \tag{6.9}$$

$$\gamma_2 = \frac{2\Delta I_R^2 \tau_0}{\pi(\tau_c + \tau_e)} \tag{6.10}$$

and

$$\tau_0 = \frac{\tau_c \tau_e}{\tau_c + \tau_e} \tag{6.11}$$

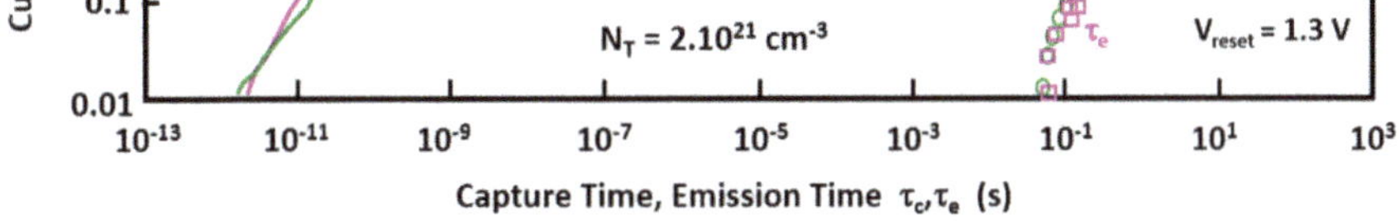

Figure 6.26. Experimental (symbols) RTS capture and emission times measured in ReRAM devices in HRS reset at V_{reset} = 1.3 V. The capture and emission times associated with the MPTAT through oxygen vacancy defects are also shown for comparison (solid lines). (Reproduced with permission from Puglisi *et al* [105]. Copyright 2015 IEEE.)

In equation (6.9), ΔI_R is the RTS current switching amplitude, while $f_{max} = (2t_s)^{-1}$ with t_s the sampling time. The model correctly explains the observation that the normalized standard deviation is initially negligibly small until the first current step occurs in the cell, at which instant σ_{IR}/I_R sharply increases [107]. The level of the normalized standard deviation does not continuously increase with time, rather it saturates to a constant value of 0.2 due to the presence of only two levels in the current distribution. This corresponds with the saturated level in the noise PSD at low frequencies or, in other words, with the plateau in the Lorentzian spectrum [107]. It has been experimentally validated that the read noise related to RTS cannot easily be suppressed by changing the composition of the oxide, as a kind of universal noise versus resistance characteristic has been found for different types of RRAM stacks [107].

Finally, a detailed experimental and modeling effort has been undertaken to understand the resistance fluctuations in an RRAM array [108]. Tail bits have been identified which show new noise behaviors called time-dependent random walk (RW) and interrupted RTN, which are illustrated in figure 6.27. RW fluctuations consist of step changes of resistance at random times and amplitudes. These have been ascribed to defect structural relaxation [108]. Intermittent RTN occurs when TLFs randomly start and stop in the RRAM cell and are thought to originate from bistable defects, switching between an inactive (off) and an active (on) state. Broadening of the resistance distribution in an array sets in already at short times and does not increase much further on—this is not due to dynamic equilibrium noise [108]. It has been shown that the number of cells with a high activity of random resistance fluctuations is only a small fraction of the array. Most of the cells display

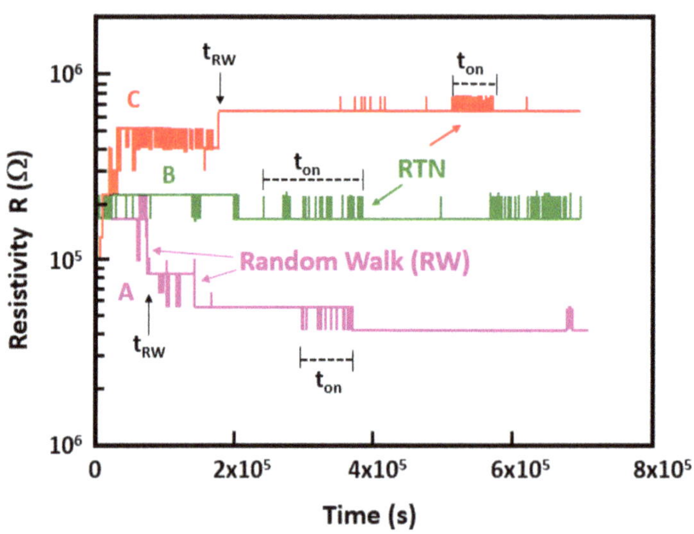

Figure 6.27. Measured resistance R as a function of time for cells A, B and C showing RW occurring at t_{RW} and interrupted RTN lasting for a time t_{on}. (Reproduced with permission from Ambrosio *et al* [108]. Copyright 2015 IEEE.)

only minor steps or no steps at all, with an equal probability for steps to a higher or a lower value. In addition, the occurrence of steps with a large amplitude becomes less probable at longer times. These RRAM fluctuations decay with time and thus cannot be explained by a stationary model of noise, such as the standard 1/*f* or RTS noise [108]. It has been shown that RW steps become increasingly less frequent as time progresses after the reset bias pulse. This can be explained by a stabilization of defects contributing to the CF in the HRS after reset, probably due to mechanical stress relaxation.

Similar interest exists for so-called PCMs [109–112], which rely on the electrically induced change between amorphous and polycrystalline phases in an active chalcogenide material, typically $Ge_2Sb_2Te_5$. Again, the resistive ratio between the two phases is high, so that switching occurs between an HRS (reset) and an LRS. In this case, the study of RTS can be useful to understand the structural stability of the amorphous phase and the conduction mechanism in the amorphous chalcogenide phase [111]. The RTS is interpreted in terms of structural fluctuations at a bistable defect which affects the local conductivity in a PP. A possible model is represented by figure 6.28 [96], which suggests that RTS is due to the interaction of a mobile species (e.g. Cu^{2+}) with the current transport through filamentary conduction. Ions (defects) move in a random potential landscape and are randomly distributed over

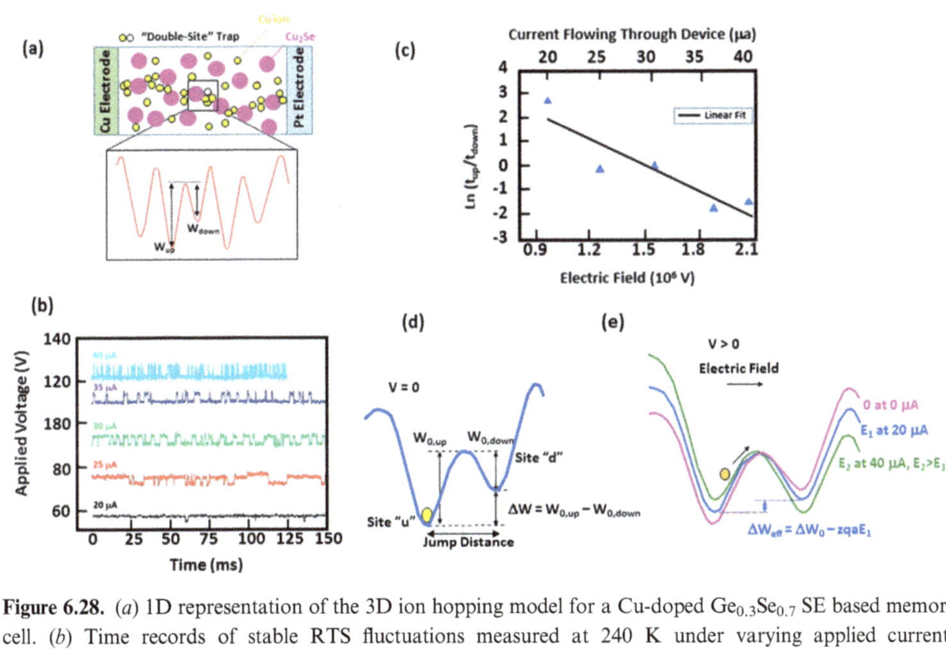

Figure 6.28. (*a*) 1D representation of the 3D ion hopping model for a Cu-doped $Ge_{0.3}Se_{0.7}$ SE based memory cell. (*b*) Time records of stable RTS fluctuations measured at 240 K under varying applied currents (i.e. potential drops) on a memory cell in a LRS of ~3 kΩ. (*c*) The ratio of the average lifetimes for the up and down state of the RTS fluctuator as a function of the average applied electric field and current flow through the memory cell. (*d*)–(*e*) Schematics of an asymmetric double-well potential for a fluctuating Cu ion without (*d*) and with (*e*) an applied electric field showing the deformation of the potential wells in that case. (Reproduced with permission from Soni *et al* [96].Copyright 2010 AIP Publishing.)

local minima (sites). The ion movement is due to activated transitions over the energy barriers separating the sites. The average fluctuation lifetimes should depend on the applied bias (or current) as this will change the potential landscape [96].

It has also been shown that $1/f$ current fluctuations can explain the statistical spread in the threshold switching [112], which is one of the major reliability concerns in PCM and other resistive memories, yielding in the end a read disturb. The delay time exhibits strong voltage dependence and an extremely low Weibull shape factor [113]. It has been shown that there exists a direct correlation between the measured statistical distribution of the delay time and the frequency spectrum of the current noise. Also the V_T distribution can be explained by the same $1/f$ model [113]. The sub-Poissonian ($\beta < 1$) statistical spread of the delay time comes from the $1/f$ dependence of the current noise.

When investigating the scaling of the LFN in PCM devices, it has been found that the noise remains $1/f$-like, scaling with 1/volume [114]. This indicates that the conduction mechanisms of $Ge_2Sb_2Te_5$ materials are bulk-like and not filamentary in nature.

6.4 RTS in CMOS imagers and CCDs

RTS has been observed in proton-irradiated CCDs for space applications [115–117] and is linked to the average leakage current per pixel. In other words, leaky pixels have a high probability to contain a trap generating RTS. This correlation is illustrated in figure 6.29 [116]. Moreover, it has been shown that the creation of RTSs is related to the displacement damage or to the non-ionizing energy loss

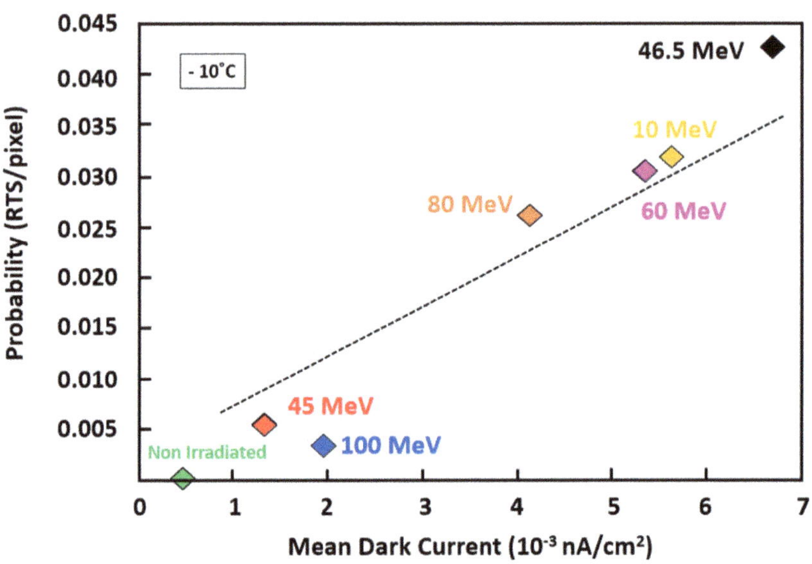

Figure 6.29. Probability of a pixel with RTS after irradiation versus the mean dark current at −10°C. (After Nuns *et al* [117].)

deposited by the irradiating particle. This strongly suggests that the traps are created in the silicon and not in the gate dielectric. A statistical approach to measure RTS in a large number of CCD pixels has been developed [117], showing tails of large amplitude signals. The high amplitude is thought to occur in high-electric-field regions, causing an enhancement of the leakage current generation by the trap. The observed activation energies below mid-gap ($E_a \sim 0.4$–0.44 eV) are in line with this interpretation. As to the origin of the defects, it has been speculated that metastable P–V pairs (or clusters) could be responsible for the anomalously high RTSs [117]. In this sense, RTSs in (irradiated) CCDs show a strong resemblance to the ones in DRAMs [1, 2, 61, 64]. This has also been concluded for the origin of stuck bits in irradiated synchronous DRAM (SDRAMs), i.e. DRAMs that are synchronized with the clock speed that the microprocessor is optimized for [118].

Radiation- [119–123] and process-induced RTSs [124–132] have also been reported in CMOS imagers. The origin of the RTS is again associated with trap levels in the silicon depletion region, giving rise to a discrete fluctuation of the photodiode dark current [121]. In addition, enhancement of the sensor temporal readout noise occurs due to the discrete fluctuation of the in-pixel source-follower channel conduction, induced by gate oxide trapping [121]. The defect metastability of generation centers in the silicon depletion region should be the main cause of the bi- or multi-level fluctuations in the dark current characteristics. A comparison between RTS in proton-irradiated CCDs and active pixel sensors demonstrates the similarity between the two types of devices, suggesting the same defect origin [133].

A new source of RTS in pinned photodiode CMOS image sensors has been pointed out [121, 122, 127], associated with metastable trapping in the bulk oxide [121]. It gives rise to slower time constants than source-follower RTS and yields an amplitude proportional to the integration time. This RTS is uncorrelated with 'hot' or blinking pixels showing a high dark current and exhibits an amplitude which is one decade higher than the shot noise in the dark current [121]. The less the pinned photodiode depletion region is in contact with the gate oxide, the fewer RTSs of this type are detected. This suggests that traps in the gate oxide close to the transfer gate are at the origin of the trapping/detrapping. The impact of ionizing radiation on this type of RTS has been studied [121, 127], showing that the number of RTS levels per pixel increases proportionally with total ionizing dose, while the average amplitude remains the same. It is not dependent on the reverse bias, with field-independent activation energy of ~0.6 eV. It was pointed out that the STI was the main contributor to this RTS [121]. An exponential distribution is observed with a slope parameter, increasing with x-ray dose [121].

RTS has also been reported in single-photon avalanche diodes (SPADs) processed in 0.8 μm MOS technology and in proton-irradiated SPADs from 0.35 μm technology [134]. The RTSs are temperature- and bias-dependent and can be correlated with the multistable behavior of the dark count rate, i.e. the average rate of registered counts without any incident light.

References

[1] Yaney D S, Lu C Y, Kohler R A, Kelly M J and Nelson J T 1987 A meta-stable leakage phenomenon in DRAM charge storage—variable hold time *Tech. Dig. Int. Electron Devices Meeting—IEDM87* (New York: IEEE) pp 336–9

[2] Restle P J, Park J W and Floyd B F 1992 DRAM variable retention time *Tech. Dig. Int. Electron Devices Meeting—IEDM92* (New York: IEEE) pp 807–10

[3] Mori Y, Ohyu K, Okomogi K and Yamada R 2005 The origin of variable retention time in DRAM *Tech. Dig. Int. Electron Devices Meeting—IEDM05* (New York: IEEE) pp 1034–7

[4] Kurata H, Otsuga K, Kotabe A, Kajiyama S, Osabe T, Sasago Y, Narumi S, Tokami K, Kamohara S and Tsuchiya O 2007 Random telegraph signal in flash memory: Its impact on scaling of multilevel flash memory beyond 90-nm node *IEEE J. Solid-State Circuits* **42** 1362–9

[5] Agostinelli M *et al* 2005 Erratic fluctuations of SRAM cache Vmin at the 90 nm process technology node *Tech. Dig. Int. Electron Devices Meeting—IEDM05* (New York: IEEE) pp 655–8

[6] Gu S, Li C W, Wang T, Lu W P, Chen K C, Ku J and Lu C-Y 2006 Read current instability arising from random telegraph noise in localized storage, multi-level flash memory *Tech. Dig. Int. Electron Devices Meeting—IEDM06* (New York: IEEE) pp 1–4

[7] Fukuda K, Shimizu Y, Amemiya K, Kamoshida M and Hu C 2007 Random telegraph noise in flash memories—model and technology scaling *Tech. Dig. Int. Electron Devices Meeting—IEDM07* (New York: IEEE) pp 169–72

[8] Tega N, Miki H, Osabe T, Kotabe A, Otsuga K, Kurata H, Kamohara S, Tokami K, Ikeda Y and Yamada R 2006 Anomalously large threshold voltage fluctuation by complex random telegraph signal in floating gate flash memory *Tech. Dig. Int. Electron Devices Meeting—IEDM06* (New York: IEEE)

[9] Abe K, Fujisawa T, Teramoto A, Watabe S, Sugawa S and Omi T 2009 Anomalous random telegraph signal extractions from a very large number of n-metal oxide semiconductor field-effect transistors using test element groups with 0.47 Hz–3.0 MHz sampling frequency *Jpn. J. Appl. Phys.* **48** 04C044

[10] Spinelli A S, Monzio Compagnoni C, Gusmeroli R, Ghidotti M and Lacaita A L 2007 RTN effects in scaled flash memory arrays *Extended Abstracts of the 2007 Int. Conf. on Solid State Devices and Materials* (Tsukuba pp 896–7

[11] Spinelli A S, Monzio Compagnoni C, Gusmeroli R, Ghidotti M and Visconti A 2008 Investigation of the random telegraph noise instability in scaled flash memory arrays *Jpn. J. Appl. Phys.* **47** 2598–601

[12] Monzio Compagnoni C, Spinelli A S, Beltrami S, Bonanomi M and Visconti A 2008 Cycling effect on the random telegraph noise instabilities of NOR and NAND flash arrays *IEEE Electron Device Lett.* **29** 941–3

[13] Monzio Compagnoni C, Chiavarone L, Calabrese M, Gusmeroli R, Ghidotti M, Lacaita A L, Spinelli A S and Visconti A 2009 Granular electron injection and random telegraph noise impact on the programming accuracy of NOR flash memories *Proc. of the 47th Annual International Reliability Physics Symp.—IRPS09* (New York: IEEE) pp 274–9

[14] Monzio Compagnoni C, Ghidotti M, Lacaita A L, Spinelli A S and Visconti A 2009 Random telegraph noise effect on the programmed threshold-voltage distribution of flash memories *IEEE Electron Device Lett.* **30** 984–6

[15] Bae S-H, Lee J-H, Kwon H-I, Ahn J-R, Om J-C, Park C H and Lee J-H 2009 The 1/*f* noise and random telegraph noise characteristics in floating-gate NAND flash memories *IEEE Trans. Electron Devices* **56** 1624–30
[16] Ma H C, Chou Y L, Chiu J P, Wang T, Ku S H, Zou N K, Chen V, Lu W P, Chen K C and Lu C-Y 2009 Program trapped-charge effect on random telegraph noise amplitude in a planar SONOS flash memory cell *IEEE Electron Device Lett.* **30** 1188–90
[17] Chiu J P, Chou Y L, Ma H C, Wang T, Ku S H, Zou N K, Chen V, Lu W P, Chen K C and Lu C-Y 2009 Program effect on random telegraph noise amplitude and its device structural dependence in SONOS Flash memory *Tech. Dig. Int. Electron Devices Meeting—IEDM09* (New York: IEEE) pp 843–6
[18] Kim T, He D, Porter R, Rivers D, Kessenich J and Goda A 2010 Comparative study of quick electron detrapping and random telegraph signal and their dependences on random discrete dopant in sub-40-nm NAND Flash memory *IEEE Electron Device Lett.* **31** 153–5
[19] Chou Y L, Chiu J P, Ma H C, Wang T, Chao Y P, Chen K C and Lu C-Y 2010 Use of random telegraph signal as internal probe to study program/erase charge lateral spread in a SONOS Flash memory *Proc. Annual International Reliability Physics Symp.—IRPS10* (New York: IEEE) pp 960–3
[20] Joe S-M, Yi J-H, Park S-K, Kwon H-I and Lee J-H 2010 Position-dependent threshold voltage variation by random telegraph noise in NAND Flash memory strings *IEEE Electron Device Lett.* **31** 635–7
[21] Janai M and Bloom I 2010 Charge gain, NBTI, and random telegraph noise in EEPROM flash memory devices *IEEE Electron Device Lett.* **31** 1038–40
[22] Janai M, Bloom I and Shur Y 2011 Charge gain, NBTI recovery and random telegraph noise in localized-trapping NVM devices *Proc. Annual International Reliability Physics Symp.—IRPS11* (New York: IEEE) pp 819–23
[23] Joe S-M, Yi J-H, Park S-K, Shin H, Park B-G, Park Y J and Lee J-H 2011 Threshold voltage fluctuation by random telegraph noise in floating gate NAND flash memory string *IEEE Trans. Electron Devices* **58** 67–73
[24] Kim T, He D, Morinville K, Sarpatwari K, Millemon B, Goda A and Kessenich J 2011 Tunnel oxide nitridation effect on the evolution of V_t instabilities (RTS/QED) and defect characterization for sub-40-nm Flash memory *IEEE Electron Device Lett.* **32** 999–1001
[25] Li F-H and Shirota R 2013 Impact of source/drain junction and cell shape on random telegraph noise in NAND Flash memory *Jpn. J. Appl. Phys.* **52** 074201
[26] Jo B-S, Kang H-J, Joe S-M, Jeong M-K, Han K-R, Park S-K, Park B-G and Lee J-H 2013 Characterization of random telegraph noise generated by process- and cycling-stress-induced traps in 26 nm NAND flash memory *Jpn. J. Appl. Phys.* **52** 04CA07
[27] Yang X, Huo Z, Jin L, Wang Z, Jiang D D, Wang Y and Liu M 2014 Program charge effect on random telegraph noise behavior in multi-level floating gate flash memory *Proc. ICSICT 2014* (New York: IEEE) pp 1637–9
[28] Yang X, Zheng Z, Wang Y, Huo Z, Jin L, Jiang D, Wang Z, Chiu S, Wu H and Liu M 2015 Gate bias dependence of complex random telegraph noise behavior in 65-nm NOR flash memory *IEEE Electron Device Lett.* **36** 26–8
[29] Resnati D, Monzio Compagnoni C, Paolucci G M, Miccoli C, Spinelli A S, Lacaita A L, Visconti A and Goda A 2015 Random telegraph noise-induced sensitivity of data retention to cell position in the programmed distribution of NAND flash memory arrays *IEEE Electron Device Lett.* **36** 678–80

[30] Yang X, Liu J, Zheng Z, Wang Y, Jiang D, Chiu S, Wu H and Liu M 2015 Impact of P/E cycling on read current fluctuation of NOR flash memory cell: a microscopic perspective based on low frequency noise analysis *Proc. Annual International Reliability Physics Symp.—IRPS15* (New York: IEEE) pp 5B.7.1–5B7.6

[31] Kim S, Lee M, Choi G-B, Lee J, Lee Y, Cho M, Ahn K-O and Kim J 2015 RTS noise reduction of 1Y-nm floating gate NAND flash memory using process optimization *Proc. Annual International Reliability Physics Symp.—IRPS15* (New York: IEEE) pp MY.8.1–MY.8.4

[32] Miccoli C, Paolucci G M, Monzio Compagnoni C, Spinelli A S and Goda A 2015 Cycling pattern and read/bake conditions dependence of random telegraph noise in decananometer NAND flash arrays *Proc. Annual International Reliability Physics Symp.—IRPS15* (New York: IEEE) pp MY.9.1–MY.9.6

[33] Tega N, Miki H, Pagette F, Frank D J, Ray A, Rooks M J, Haensch W and Torii K 2009 Increasing threshold voltage variation due to random telegraph noise in FETs as gate lengths scale to 20 nm *Symp. on VLSI Techn. Dig. of Techn. Papers* (New York: IEEE) pp 50–1

[34] Takeuchi K, Nagumo T, Yokogawa S, Imai K and Hayashi Y 2009 Single-charge-based modeling of transistor characteristics fluctuations based on statistical measurement of RTN amplitude *Symp. on VLSI Techn. Dig. of Techn. Papers* (New York: IEEE) pp 54–5

[35] Wang R-V, Lee Y-H, Lu Y-L R, McMahon W, Hu S and Ghetti A 2009 Shallow trench isolation edge effect on random telegraph signal noise and implications for Flash memory *IEEE Trans. Electron Devices* **56** 2107–13

[36] Li S-R, McMahon W, Lu Y-L R and Lee Y-H 2008 RTS noise characterization in flash cells *IEEE Electron Device Lett.* **29** 106–8

[37] Ghetti A, Monzio Compagnoni C, Biancardi F, Lacaita A L, Beltrami S, Chiavarone L, Spinelli A S and Visconti A 2008 Scaling trends for random telegraph noise in deca-nanometer flash memories *Tech. Dig. Int. Electron Devices Meeting—IEDM08* (New York: IEEE)

[38] Ghetti A, Monzio Compagnoni C, Spinelli A S and Visconti A 2009 Comprehensive analysis of random telegraph noise instability and its scaling in deca-nanometer flash memories *IEEE Trans. Electron Devices* **56** 1746–52

[39] Monzio Compagnoni C, Gusmeroli R, Spinelli A S, Lacaita A L, Bonanomi M and Visconti A 2008 Statistical model for random telegraph noise in flash memories *IEEE Trans. Electron Devices* **55** 388–95

[40] Ghetti A, Bonanomi M, Spinelli A S, Lacaita A L and Visconti A 2008 Physical modeling of single-trap RTS statistical distribution in flash memories *Proc. 46th Annual International Reliability Physics Symp.—IRPS08* (New York: IEEE) pp 610–5

[41] Chimenton A, Zambelli C and Olivo P 2009 A statistical model for erratic erase based on an automated random telegraph signal characterization technique *Proc. 47th Annual International Reliability Physics Symp.—IRPS09* (New York: IEEE) pp 896–901

[42] Asenov A, Balasubramanian R, Brown A R and Davies J H 2003 RTS amplitudes in decananometer MOSFETs: 3D simulation study *IEEE Trans. Electron Devices* **50** 839–45

[43] Ghetti A, Amoroso S M, Mauri A and Compagnoni C M 2012 Impact of nonuniform doping on random telegraph noise in flash memory devices *IEEE Trans. Electron Devices* **59** 309–15

[44] Amoroso S M, Ghetti A, Brown A R, Mauri A, Monzio Compagnoni C and Asenov A 2012 Impact of cell shape on random telegraph noise in decananometer flash memories *IEEE Trans. Electron Devices* **59** 2774–9

[45] Jeong M K, Joe S-M, Seo C-S, Han K-R, Choi E, Park S-K and Lee J-H 2012 Analysis of random telegraph noise and low frequency noise properties in 3D stacked NAND flash memory with tube-type poly-Si channel structure *Symp. on VLSI Technol. Dig. of Techn. Papers* (New York: IEEE) pp 55–6
[46] Kang D, Lee C, Hur S, Song D and Choi J-H 2014 A new approach for trap analysis of vertical NAND flash cell using RTN characteristics *Tech. Dig. Int. Electron. Devices Meeting—IEDM14* (New York: IEEE) pp 367–70
[47] Degraeve R *et al* 2011 Statistical characterization of current paths in narrow poly-Si channels *Tech. Dig. Int. Electron. Devices Meeting—IEDM11* (New York: IEEE) pp 287–90
[48] de Andrade M G C, Luque Toledano M, Fourati F, Degraeve R, Martino J A, Claeys C, Simoen E, Van den Bosch G and Van Houdt J 2013 RTN assessment of traps in polysilicon cylindrical vertical FETs for NVM application *Microelectron. Eng.* **109** 105–8
[49] Toledano-Luque M *et al* 2013 Statistical spectroscopy of switching traps in deeply scaled vertical poly-Si channel for 3D memories *Tech. Dig. Int. Electron. Devices Meeting—IEDM13* (New York: IEEE) pp 562–5
[50] Lee K H, Degraeve R, Toledano-Luque M, Arreghini A, Breuil L, Blomme P, Van den Bosch G and Van Houdt J 2015 Assessment of tunnel oxide and poly-Si channel traps in 3D SONOS memory before and after P/E cycling *Microelecetron. Eng.* **147** 45–50
[51] Simoen E, Cretu B, Fang W, Aoulaiche M, Routoure J-M, Carin R, Luo J, Zhao C and Claeys C 2016 Low frequency noise spectroscopy of bulk and border traps in nanoscale devices *Solid State Phenom.* **242** 449–58
[52] Tega N, Miki H, Yamaoka M, Kume H, Mine T, Ishida T, Mori Y, Yamada R and Torii K 2008 Impact of threshold voltage fluctuation due to random telegraph noise in scaled-down SRAM *Proc. of the 46th Annual International Reliability Physics Symp.—IRPS08* (New York: IEEE) pp 541–6
[53] Yamaoka M, Miki H, Bansal A, Frank D J, Leobandung E and Torii K 2011 Evaluation methodology for random telegraph noise effects in SRAM arrays *Tech. Dig. Int. Electron Devices Meeting—IEDM11* (New York: IEEE) pp 745–8
[54] Toh S O, Tsukamoto Y, Guo Z, Jones L, Liu T-J K and Nikolić B 2009 Impact of random telegraph signals on V_{min} in 45nm SRAM *Tech. Dig. Int. Electron Devices Meeting—IEDM11* (New York: IEEE) pp 767–70
[55] Takeuchi K, Nagumo T, Takeda K, Asayama S, Yokogawa S, Imai K and Hayashi Y 2010 Direct observation of RTN-induced SRAM failure by accelerated testing and its application to product reliability assessment *Symp. on VLSI Technol. Dig. of Tech. Papers* (New York: IEEE) pp 189–90
[56] Toh S O, Liu T-J K and Nikolić B 2011 Impact of random telegraph signaling noise on SRAM stability *Symp. on VLSI Technol. Dig. of Tech. Papers* (New York: IEEE) pp 204–5
[57] Zimmer B, Thomas O, Toh S O, Vincent T, Asanović K and Nikolić B 2014 Joint impact of random variations and RTN on dynamic writeability in bulk and FDSOI SRAM *IEEE Proc.* (New York: IEEE) pp 98–101
[58] Qiu H *et al* 2015 Impact of random telegraph noise on write stability in silicon-on-thin-BOX (SOTB) SRAM cells at low supply voltage in sub-0.4 V regime *Symp. on VLSI Technol. Dig. of Techn. Papers* (New York: IEEE) pp T38–T39
[59] Mori Y, Takeda K and Yamada R 2010 Random telegraph noise of junction leakage current in submicron devices *J. Appl. Phys.* **107** 014509

[60] Umeda T, Okonogi K, Ohyu K, Tsukada S, Hamada K, Fujieda S and Mochizuki Y 2006 Single silicon vacancy-oxygen complex defect and variable retention time phenomenon in dynamic random access memories *Appl. Phys. Lett.* **88** 253504

[61] Ohyu K, Umeda T, Okonogi K, Tsukada S, Hidaka M, Fujieda S and Mochizuki Y 2006 Quantitative identification for the physical origin of variable retention time: A vacancy-oxygen complex defect model *Tech. Dig. Int. Electron Devices Meeting—IEDM06* (New York: IEEE) pp 1–4

[62] Mori Y, Shima A, Takeda K and Yamada R 2012 Experimental evidence for involvement of interface states in random telegraph noise in junction leakage current of metal–oxide–semiconductor field-effect transistor *Jpn. J. Appl. Phys.* **51** 1–4

[63] Uren M J, Kirton M J and Collins S 1988 Anomalous telegraph noise in small-area silicon metal–oxide–semiconductor field-effect transistors *Phys. Rev.* B **37** 8346–50

[64] Kim H, Kim K, Oh T-K, Cha S-Y, Hong S-J, Park S-W and Shin H 2009 RTS-like fluctuation in gate induced drain leakage current of saddle-fin type DRAM cell transistor *Tech. Dig. Int. Electron Devices Meeting—IEDM09* (New York: IEEE) pp 271–4

[65] Oh B, Cho H-J, Kim H, Son Y, Kang T, Park S, Jang S, Lee J-H and Shin H 2011 Characterization of an oxide trap leading to random telegraph noise in gate-induced drain leakage current of DRAM cell transistors *IEEE Trans. Electron Devices* **58** 1741–7

[66] Kim H, Oh B, Son Y, Kim K, Cha S-Y, Jeong J-G, Hong S-J and Shin H 2011 Characterization of the variable retention time in dynamic random access memory *IEEE Trans. Electron Devices* **58** 2952–8

[67] Ito K, Matsumoto T, Nishizawa S, Sunagawa H, Kobayashi K and Onodera H 2011 The impact of RTN on performance fluctuation in CMOS logic circuits *Proc. Annual International Reliability Physics Symp.—IRPS11* (New York: IEEE) pp 710–3

[68] Terai M, Sakotsubo Y, Saito Y, Kotsuji S and Hada H 2009 Effect of bottom electrode of ReRAM with Ta_2O_5/TiO_2 stack on RTN and retention *Tech. Dig. Int. Electron Devices Meeting—IEDM09* (New York: IEEE) pp 775–8

[69] Terai M, Sakotsubo Y, Saito Y, Kotsuji S and Hada H 2010 Memory-state dependence of random telegraph noise of Ta_2O_5/TiO_2 stack ReRAM *IEEE Electron Device Lett.* **31** 1302–4

[70] Tseng Y H, Shen W C, Huang C-E, Lin C J and King Y-C 2010 Electron trapping effect on the switching behavior of contact RRAM devices through random telegraph noise analysis *Tech. Dig. Int. Electron Devices Meeting—IEDM10* (New York: IEEE) pp 636–9

[71] Tseng Y H, Shen W C and Lin C J 2012 Modeling of electron conduction in contact resistive random access memory devices as random telegraph noise *J. Appl. Phys.* **111** 1–5

[72] Chang L-S, Lin C J and King Y-C 2014 Temperature dependent characteristics of the random telegraph noise on contact resistive random access memory *Jpn. J. Appl. Phys.* **53** 1–4

[73] Lee M-S, Lee J-K, Hwang H-S, Shin H-C, Park B-G, Park Y-J and Lee J-H 2011 Conduction mechanism and low frequency noise analysis in $Al/Pr_{0.7}Ca_{0.3}MnO_3$ for bipolar resistive switching *Jpn. J. Appl. Phys.* **50** 011501

[74] Lee S B *et al* 2009 Large $1/f$ noise of unipolar resistance switching and its percolating nature *Appl. Phys. Lett.* **95** 1–3

[75] Ielmini D, Nardi F and Cagli C 2010 Resistance-dependent amplitude of random telegraph-signal noise in resistive switching memories *Appl. Phys. Lett.* **96** 1–3

[76] Lee J-K, Jeong H Y, Cho I-T, Lee J Y, Choi S-Y, Kwon H-I and Lee J-H 2010 Conduction and low-frequency noise analysis in $Al/\alpha\text{-}TiO_x/Al$ bipolar switching resistance random access memory devices *IEEE Electron Device Lett.* **31** 603–5

[77] Lee J-K, Lee J-W, Park J, Chung S-W, Roh J S, Hong S-J, Cho I, Kwon H-I and Lee J-H 2011 Extraction of trap location and energy from random telegraph noise in amorphous TiO_x resistance random access memories *Appl. Phys. Lett.* **98** 1–3

[78] Lee J-K, Cho I-T, Kwon H-I, Hwang C S, Park C H and Lee J-H 2012 Relationship between conduction mechanism and low-frequency noise in polycrystalline-TiO_x-based resistive switching memory devices *IEEE Electron Device Lett.* **33** 1063–5

[79] Kim J H, Lee A R, Bae Y C, Baek K H, Im H S and Hong J P 2014 Observation of bias-dependent noise sources in a TiO_x/TiO_y bipolar resistive switching frame *J. Appl. Phys.* **104** 1–4

[80] Lee D, Lee J, Jo M, Park J, Siddik M and Hwang H 2011 Noise-analysis-based model of filamentary switching ReRAM with ZrO_x/HfO_x stacks *IEEE Electron Device Lett.* **32** 964–6

[81] Zhang Y, Wu H, Wu M, Deng N, Yu Z, Zhang J and Qian H 2014 Random telegraph noise analysis in AlO_x/WO_y resistive swithcing memories *Appl. Phys. Lett.* **104** 1–4

[82] Fang Z, Yu H Y, Chroboczek J A, Ghibaudo G, Buckley J, De Salvo B, Li X and Kwong D L 2012 Low-frequency noise in oxide-based (TiN/HfO_x/Pt) resistive random access memory cells *IEEE Trans. Electron Devices* **59** 850–3

[83] Puglisi F M, Pavan P, Padovani A, Larcher L and Bersurker G 2012 Random telegraph signal noise properties of HfO_x RRAM in high resistive state *Proc. European Solid-State Device Research Conf.—ESSDERC 2012* (New York: IEEE) pp 274–7

[84] Puglisi F M, Pavan P, Padovani A, Larcher L and Bersuker G 2013 RTS noise characterization of HfO_x RRAM in high resistive state *Solid-State Electron.* **84** 160–6

[85] Raghavan N, Degraeve R, Goux L, Fantini A, Wouters D J, Groeseneken G and Jurczak M 2013 RTN insight to filamentary instability and disturb immunity in ultra-low power switching HfO_x and AlO_x RRAM *Symp. on VLSI Technol. Dig. of Tech. Papers* (New York: IEEE) pp T164–T165

[86] Yu S, Guan X and Wong H-S P 2011 On the stochastic nature of resistive switching in metal oxide RRAM: physical modeling, Monte Carlo simulation, and experimental characterization *Tech. Dig. Int. Electron Devices Meeting—IEDM11* (New York: IEEE) pp 413–6

[87] Yu S, Jeyasingh R, Wu Y and Wong H-S P 2011 Understanding the conduction and switching mechanism of metal oxide RRAM through low frequency noise and AC conductance measurement and analysis *Tech. Dig. Int. Electron Devices Meeting—IEDM11* (New York: IEEE) pp 275–8

[88] Gonzalez M B, Rafí J M, Beldarrain O, Zabala M and Campabadal F 2014 Analysis of the switching variability in Ni/HfO_2-based RRAM devices *IEEE Trans. Device and Mater. Reliab.* **14** 769–72

[89] Chung Y T, Liu Y H, Su P C, Cheng Y H, Wang T and Chen M C 2014 Investigation of random telegraph noise amplitudes in hafnium oxide resistive memory devices *Proc. Annual International Reliability Physics Symp. 14* (New York: IEEE) pp MY2.1–MY.2.5

[90] Gonzalez M B, Martin-Martinez J, Rodríguez R, Acero M C, Nafría M, Campabadal F and Aymerich X 2015 Dedicated random telegraph noise characterization of Ni/HfO_2-based RRAM devices *Microelectron. Eng.* **147** 59–62

[91] Maestro M, Diaz J, Crespo-Yepes A, Gonzalez M B, Martin-Martinez J, Rodriguez R, Nafria M, Campabadal F and Aymerich X 2016 New high resolution random telegraph noise (RTN) characterization method for resistive RAM *Solid-State Electron.* **115** 140–5

[92] Ambrogio S, Balatti S, Cubeta A, Calderoni A, Ramaswamy N and Ielmini D 2013 Understanding switching variability and random telegraph noise in resistive RAM *Tech. Dig. Int. Electron Devices Meeting—IEDM13* (New York: IEEE) pp 782–5
[93] Chang L-S, Huang C-Y, Tseng Y-H, King Y-C and Lin C-Y 2013 Temperature sensing scheme through random telegraph noise in contact RRAM *IEEE Electron Device Lett.* **34** 12–4
[94] Balatti S, Ambrogio S, Bubeta A, Calderoni A, Ramaswamy N and Ielmini D 2014 Voltage-dependent random telegraph noise (RTN) in HfO_x resistive RAM *Proc. Annual International Reliability Physics Symp.—IRPS14* (New York: IEEE) pp MY.4.1–MY.4.6
[95] Ambrogio S, Balatti S, Cubeta A, Calderoni A, Ramaswamy N and Ielmini D 2014 Statistical fluctuations in HfO_x resistive-switching memory. Part II: Random telegraph noise *IEEE Trans. Electron Devices* **61** 2920–7
[96] Soni R, Meuffels P, Petraru A, Weides M, Kügeler C, Waser R and Kohlstedt H 2010 Probing Cu doped $Ge_{0.3}Se_{0.7}$ based resistance switching memory devices with random telegraph noise *J. Appl. Phys.* **107** 024517
[97] Ralls K S and Buhrman R A 1991 Microscopic study of 1/*f* noise in metal nanobridges, *Phys. Rev.* B **44** 5800–17
[98] Puglisi F M and Pavan P 2013 RTN analysis with FHMM as a tool for multi-trap characterization in HfO_x RRAM *Proc. IEEE Int. Conf. Electron Devices Solid-State Circuits (EDSCC)* (New York: IEEE) pp 1–2
[99] Puglisi F M, Larcher L, Pavan P, Padovani A and Bersuker G 2014 Instability of HfO_2 RRAM devices: comparing RTN and cycling variability *Proc. Annual International Reliability Physics Symp.—IRPS14* (New York: IEEE) pp MY.5.1–MY.5.5
[100] Puglisi F M, Pavan P, Padovani A and Larcher L 2014 A study on HfO_2 RRAM in HRS based on *I–V* and RTN analysis *Solid-State. Electron* **102** 69–75
[101] Puglisi F M, Larcher L, Padovani A and Pavan P 2015 A complete statistical investigation of RTN in HfO_2-based RRAM in high resistive state *IEEE Trans. Electron Devices* **62** 2606–13
[102] Puglisi F M, Pavan P, Larcher L and Padovani A 2014 Analysis of RTN and cycling variability in HfO_2 RRAM devices in LRS *Proc. IEEE* (New York: IEEE) pp 246–9
[103] Puglisi F M, Pavan P, Larcher L and Padovani A 2015 Statistical analysis of random telegraph noise in HfO_2-based RRAM devices in LRS *Solid-State Electron* **113** 132–7
[104] Degraeve R, Fantini A, Raghavan N, Goux L, Clima S, Govoreanu B, Belmonte A, Linten D and Jurczak M 2015 Causes and consequences of the stochastic aspect of filamentary RRAM *Microelectron. Eng.* **147** 171–5
[105] Puglisi F M, Pavan P, Vandelli L, Padovani A, Bertocchi M and Larcher L 2015 A microscopic physical description of RTN current fluctuations in HfO_x RRAM *Proc. Annual International Reliability Physics Symp.—IRPS15* (New York: IEEE) 5B.5.1–5B5.6
[106] Ambrogio S, Balatti S, McCaffrey V, Wang D and Ielmini D 2014 Impact of low-frequency noise on read distributions of resistive switching memory (RRAM) *Tech. Dig. Int. Electron Devices Meeting—IEDM14* (New York: IEEE) pp 363–6
[107] Ambrogio S, Balatti S, McCaffrey V, Wang D C and Ielmini D 2015 Noise-induced resistance broadening in resistive switching memory. Part I: Intrinsic cell behavior *IEEE Trans. Electron Devices* **62** 3805–11
[108] Ambrogio S, Balatti S, McCaffrey V, Wang D C and Ielmini D 2015 Noise-induced resistance broadening in resistive switching memory. Part II: Array statistics *IEEE Trans. Electron Devices* **62** 3812–9

[109] Fantini P, Pirovani A, Ventrice D and Redaelli A 2006 Experimental investigation of transport properties in chalcogenide materials through 1/*f* noise analysis *Appl. Phys. Lett.* **88** 263506

[110] Betti Beneventi G, Calderoni A, Fantini P, Larcher L and Pavan P 2009 Analytical model for low-frequency noise in amorphous chalcogenide-based phase-change memory devices *J. Appl. Phys.* **106** 054506

[111] Fugazza D, Ielmini D, Lavizzari S and Lacaita A L 2009 Distributed-Poole–Frenkel modeling of anomalous resistance scaling and fluctuations in phase-change memory (PCM) devices *Tech. Dig. Int. Electron Devices Meeting—IEDM09* (New York: IEEE) pp 723–6

[112] Fugazza D, Ielmini D, Lavizzari S and Lacaita A L 2010 Random telegraph signal noise in phase change memory devices *Proc. Annual International Reliability Physics Symp.—IRPS10* (New York: IEEE) pp 743–9

[113] Lavizzari S, Sharna D and Ielmini D 2010 Threshold-switching delay controlled by 1/*f* current fluctuations in phase-change memory devices *IEEE Trans. Electron Devices* **57** 1047–54

[114] Betti Beneventi G, Ferro M and Fantini P 2012 1/*f* noise in 45-nm RESET-state phase-change memory devices: characterization, impact on memory readout operation, and scaling perspectives *IEEE Electron Device Lett.* **33** 1559–61

[115] Hopkins I H and Hopkinson G R 1993 Random telegraph signals from proton-irradiated CCDs *IEEE Trans. Nucl. Sci.* **40** 1567–74

[116] Hopkins I H and Hopkinson G R 1995 Further measurements of random telegraph signals in proton irradiated CCDs *IEEE Trans. Nucl. Sci.* **42** 2074–81

[117] Nuns T, Quadri G, David J-P, Gilard O and Boudou N 2006 Measurements of random telegraph signal in CCDs irradiated with protons and neutrons *IEEE Trans. Nucl. Sci.* **53** 1764–71

[118] Chugg A M, Burnell A J, Duncan P H, Parker S and Ward J J 2009 The random telegraph signal behavior of intermittently stuck bits in SDRAMs *IEEE Trans. Nucl. Sci.* **56** 3057–64

[119] Bogaerts J, Dierickx B and Mertens R 2002 Random telegraph signals in a radiation-hardened CMOS active pixel sensor *IEEE Trans. Nucl. Sci.* **49** 249–57

[120] Martin-Gonthier P and Magnan P 2009 RTS noise impact in CMOS image sensors readout circuit *Proc. 16th IEEE Int. Conf. on Electronics, Circuits, and Systems (ICECS2009)* (New York: IEEE) 928–31

[121] Goiffon V, Hopkinson G R, Magnan P, Bernard F, Rolland G and Saint-Pé O 2009 Multilevel RTS in proton irradiated CMOS image sensors manufactured in a deep submicron technology *IEEE Trans. Nucl. Sci.* **56** 2132–41

[122] Goiffon V, Magnan P, Martin-Gonthier P, Virmontois C and Gaillardin M 2011 Evidence of a novel source of random telegraph signal in CMOS image sensors *IEEE Electron Device Lett.* **32** 773–5

[123] Martin E, Nuns T, Virmontois C, David J-P and Gilard O 2013 Proton and γ-rays irradiation-induced dark current random telegraph signals in a 0.18-μm CMOS image sensor *IEEE Trans. Nucl. Sci.* **60** 2503–10

[124] Martin-Gonthier P, Havard E and Magnan P 2010 Custom transistor layout design techniques for random telegraph signal noise reduction in CMOS image sensors *Electron. Lett.* **46** 1323–4

[125] Martin-Gonthier P, Goiffon V and Magnan P 2012 In-pixel source follower transistor RTS noise behavior under ionizing radiation in CMOS image sensors *IEEE Trans. Electron Devices* **59** 1686–92

[126] Woo J-M, park H-H, Hong S-M, Chung I-Y, Min H S and Park Y J 2009 Statistical noise analysis of CMOS image sensors in dark condition *IEEE Trans. Electron Devices* **56** 2481–8

[127] Goiffon V, Virmontois C and Magnan P 2011 Investigation of dark current random telegraph signal in pinned photodiode CMOS image sensors *Tech. Dig. Int. Electron Devices Meeting—IEDM11* (New York: IEEE) pp 183–6

[128] Yonezawa A, Teramoto A, Obara T, Kuroda R, Sugawa S and Ohmi T 2013 The study of time constant analysis in random telegraph noise at the subthreshold voltage region *Proc. Annual International Reliability Physics Symp.—RPS13* (New York: IEEE) pp XT11.1–XT116

[129] Kwon H-M, Han I-S, Kwon S-K, Jang J-H, Kwak H-Y, Choi W-I, Ha M-L, Lee J-I, Hwang H and Lee H-D 2013 Effects of high-pressure annealing on random telegraph signal noise characteristic of source follower in CMOS image sensor *IEEE Electron Device Lett.* **34** 190–2

[130] Kuroda R, Yonezawa A, Teramoto A, Li T-L, Tochigi Y and Sugawa S 2013 A statistical evaluation of random telegraph noise of in-pixel source follower equivalent surface and buried channel transistors *IEEE Trans. Electron Devices* **60** 3555–61

[131] Liqiang H, Suying Y, Jiangtao X and Chao X 2013 Characteristics of random telegraph signal noise in time delay integration CMOS image sensor *Microelecton. Reliab.* **53** 400–4

[132] Pardo F, Boluda J A and Vagara F 2015 Random telegraph signal transients in active logarithmic continuous-wave vision sensors *Solid-State Electron* **114** 111–4

[133] Hopkinson G R, Goiffon V and Mohammadzadeh A 2008 Random telegraph signals in proton irradiated CCDs and APS *IEEE Trans. Nucl. Sci.* **55** 2197–204

[134] Karami M A, Carrara L, Niclass C, Fishburn M and Chabon E 2010 RTS noise characterization in single-photon avalanche diodes *IEEE Electron Device Lett.* **31** 692–4

Chapter 7

General conclusions

After more than 30 years of study, a high level of understanding of RTSs in semiconductor devices has been reached. Good insight into the underlying physics has been obtained, enabling the qualitative and quantitative modeling of their effect on device and circuit operation. At the same time, powerful measurement and analysis techniques have become available in order to collect sufficient, statistically relevant data on a large set of devices. The scaling of the technology nodes and the device feature size has led to further refinements of the modeling and extraction methodology of the RTS parameters, including quantum confinement and quantization effects, the presence of PV and short-channel effects. This goes hand in hand with the development of powerful device and circuit simulation tools, enabling the assessment of the dynamic variability which is associated with RTS.

It has become clear that RTS fluctuations in the drain or gate current can be generated by two basic mechanisms: the first based on carrier trapping by a defect which undergoes a configurational transition upon capture of a charge carrier and the second on metastable defects undergoing a vibrationally (thermally) driven switching between configurations, which changes the energy level and, hence, the carrier GR rate. While the first mechanism is believed to explain the thermally activated carrier capture and emission for the case of channel RTSs and possibly process-induced gate current signals, the second mechanism is clearly at stake for devices suffering from excessive bulk leakage current, such as DRAMs and proton-irradiated CCDs. It is probably also responsible for most of the TLFs in degraded (stressed, irradiated, etc) gate oxides.

In spite of the progress made since 1984, there are still some issues warranting further research. Hot items are the impact of the gate stack processing on the occurrence of RTSs (high-κ, metal gate, cap layers, etc). On the one hand, the replacement of high-quality SiO_2 by a more defective high-κ material certainly increases the probability for RTS (and higher $1/f$ noise). Also, replacing thin SiO_2 by a thicker high-κ layer leads to a gate further away from the traps and so to a higher

RTS amplitude. This can be partly counteracted by replacing a polysilicon gate by a metal one, at the same time allowing the reduction of the substrate doping concentration and the associated PV. The same goes for replacing a single-gate bulk architecture by a multi-gate SOI one. It is felt that all these factors play a role in the RTS behavior and associated 1/*f* noise and require further study. Another factor which has not been investigated yet is the effect of WF fluctuations, present in state-of-the-art metal gates with cap layers on the RTS amplitude and the LFN in general. Like RDF or fixed oxide fluctuations, this could amplify RTS and become a problem in ultra-thin oxide sub-14 nm CMOS technologies. In addition, little is known about RTS in novel device concepts (TFETs, NWFETs, etc) or in devices based on alternative substrates (Ge, III–V, graphene). It is known for example that depositing the same high-κ layer on germanium compared with silicon leads to a different quality oxide, as Ge atoms have the tendency to diffuse in the high-κ layer. This will create more active traps and probably different types of RTSs. While the underlying physics is the same its relative effect may be different, based on different material constants (band gap, dielectric constant, etc).

Another concern is the fact that the integration of Ge or III–V materials on a silicon platform leads to the introduction of extended defects (threading dislocations, stacking faults, twins and anti-phase boundaries, etc) in the active device layers, which operate as GR centers. Considering that the volume of the devices is on the order of 10^3 nm^3, it is clear that there is either one or no extended defect present. The same considerations hold for processing-induced (point) defects. In other words, the GR centers in the semiconductor body will also act as single GR centers, giving rise to RTS in the time domain. This implies that there is already a need for methods to distinguish semiconductor RTSs from gate oxide ones. At the same time, and as for the gate oxide traps, the investigation of channel RTSs provides a unique opportunity for defect spectroscopy in semiconductor materials. This goes even beyond the determination of the energy level and capture cross section, as it should be possible to determine the location of the trap along the channel by carefully analyzing the dependence of the RTS parameters on the operation conditions (gate and drain bias, substrate bias, etc). While noise and RTS research on novel 2D materials (MoS_2, $MoSe_2$, etc) has only just begun, it is expected that RTS can also provide valuable information on the active defects and transport mechanisms of these materials.

A final tendency dawning on the horizon is the use of *ab initio* DFT calculations to derive a possible defect structure and identity in gate dielectrics. The initial results are very promising and it is felt that this is the final piece of the puzzle in order to make RTS a truly spectroscopic tool for defect identification in semiconductor materials and devices.

www.ingramcontent.com/pod-product-compliance
Ingram Content Group UK Ltd.
Pitfield, Milton Keynes, MK11 3LW, UK
UKHW051854100926
8556IPUK00005B/627

9 780750 31273